VIRTUAL
SOCIETY

The Metaverse and the New Frontiers of Human Experience

VIRTUAL SOCIETY

HERMAN NARULA

BUSINESS

PENGUIN BUSINESS

UK | USA | Canada | Ireland | Australia
India | New Zealand | South Africa

Penguin Business is part of the Penguin Random House group of companies
whose addresses can be found at global.penguinrandomhouse.com.

First published in the United States of America by Currency, an imprint of Random House,
a division of Penguin Random House LLC 2022
First published in Great Britain by Penguin Business 2022
001

Printed and bound in Great Britain by Clays Ltd, Elcograf S.p.A.

The authorized representative in the EEA is Penguin Random House Ireland,
Morrison Chambers, 32 Nassau Street, Dublin D02 YH68

A CIP catalogue record for this book is available from the British Library

ISBN: 978–0–241–61659–8

Follow us on LinkedIn: https://www.linkedin.com/company/penguin-connect/

www.greenpenguin.co.uk

Penguin Random House is committed to a
sustainable future for our business, our readers
and our planet. This book is made from Forest
Stewardship Council® certified paper.

To Elsa, for whom all worlds open

Contents

Introduction

One day this book will be read by a person without a body.

This prediction will almost certainly come true before the end of the twenty-first century, perhaps even by 2040. Consider: We already know with some confidence that the mind is a machine that processes information. Connecting it to a computer—one with the capacity to simulate an entire world—is a fully plausible outcome and, I would argue, an inevitable one. Through developments in the field of quantum metrology, we are already able to create sensors that can listen to the electromagnetic "whispers" of the activity of clusters of neurons. Biocompatible carbon nanotubes, which are immensely strong and immensely conductive, hold promise as "neural laces": the building blocks of connections to individual neurons. If a brain can connect to a computer—whatever that eventually means—and if the computers of the future can create worlds that are as detailed as or more detailed than the one we know today, then surely a life mediated by the limits of a physical body will one day seem a pale shadow of the life of the unfettered mind.

This theoretical reader without a body—what some would call a "post-human"—will be able to access and process information in ways we can't even begin to understand today. What might it feel like to consume this text as information directly delivered to your mind in a digital reality? Maybe comprehension and integration of the book's ideas will happen instantly or nonlinearly, the hundreds of concepts forming into structures in the post-human brain at the speed of an exploding firework. Perhaps these disembodied readers will ingest this book via new senses that haven't yet been invented, or will revel in a synesthetic poetry of sounds, smells, and touch as they explore concepts with a fidelity and detail impossible through the bodily senses alone. The adventures we read about in history, science fiction, or fantasy could become the actual embodied reality of a post-human society. This post-human will live a thousand parallel lives in realities we can barely fathom.

As technology and its applications continue to improve and evolve, we are approaching a new epoch in human history, one in which the possibilities of our lives will diverge from the limits of our bodies. At that point, the world of ideas will generate actual worlds that we can inhabit: constructed realities that will exist in conversation with the physical world. Achieving this vision requires no new physics, only the inevitable waves of improvement and upgrade that we've become so good at as a species. The physical tools we've devised have transformed the Earth while enhancing our lives upon it. Our many cultural technologies—the myths and stories and rituals that have grown up alongside our tools—give form and meaning to our innovations. Reshaping our environments, for good or ill, has been key to the survival of our species since prehistory. We have always used our imaginations in tandem with our hands to explore new worlds while expanding our own. That is the human impulse, and this dynamic will persist even as our "hands" will increasingly become just a figure of speech.

This vision of a virtual future might strike you as dystopian. Maybe you envision humans being reduced to rows of bloodless, pulsating brains in jars, or you worry that technological change is happening too fast. Maybe you fear that our own world may devolve into waste and chaos as we escape into cyberspace. Perhaps to you the prospect of a life mediated by machines seems like one in which we'll be deprived of our essential humanity.

But I'd challenge you to set aside your preconceptions and consider the following: Throughout the history of our species, we humans have always imagined other, better futures for ourselves, intangible worlds that we expect to be more fulfilling and experientially rich than our daily lives. Our ability to visualize and believe in these futures is itself a cultural technology, one that we use to improve our experiences of life and reality. Depictions of the afterlife, created by artists for millennia, aren't just manifestations of religious devotion: They are extensions of an ongoing human impulse to instantiate the intangible, to visualize ideal worlds and thus make them real. We have always wanted to see, feel, and understand more than we do, and in pursuit of these goals we have consistently tried to transcend the limits imposed on us by biology and geology, and extend ourselves into potential worlds mediated only by our minds.

This important and necessary social transformation does not require the direct connection of the brain to a machine. While brain-computer interfaces will mark the most dramatic final stage in this progression, the next stage of this process will see us focusing our social and cultural attentions into a series of constructed digital realities. Today, these simulations are known as virtual worlds: embodied, three-dimensional digital spaces in which people interact via avatars. These complex graphical environments, previously thought to be the province of video games and entertainment alone, are now evolving into something much more. A "metaverse" of virtual worlds extending into every aspect of our culture is starting to emerge, presenting new economic and social opportunities that are comparable in scope to the disruption caused by the internet. Many people have characterized the metaverse and virtual worlds as a fad, or simply an evolution of video games. I believe this limited framework is based on a fundamental misunderstanding of why humans create other realities and how we seek fulfillment.

Confusion over what the metaverse is will generate broken, inconsistent models of what it is for—which, in turn, will ultimately lead to wasteful allocation of capital, ineffective attempts at regulation, and a magnification of the negative aspects of this disruption. My aim with this book is to help prevent these outcomes. In the pages that follow I will offer a new way to understand this massive transition toward a virtual society— a transition that, if we manage it carefully and learn from the mistakes of the first age of the internet, will offer humankind incomparable new dimensions of freedom.

Within decades, the worlds we create out of pixels and populate with avatars will come to matter intensely to hundreds of millions of people; the value found within these worlds will be co-created by an ever larger cross-section of society. Eventually, these worlds may well turn out to be indistinguishable from our own. When this moment arrives, it won't be a dark day for humanity: It will be the ultimate realization of an ingenious exploratory impulse that is as old as the human race. Rather than representing society's demoralizing descent into technological escapism, the emergence of virtual society will mark the beginning of an era in which we will explore new, positive frontiers in psychological fulfillment and mental health; recenter our economy and modes of education around individual needs; forge remarkable new communities built on shared interests and experiences; and bring about a world more humane than the one we are afraid to leave behind.

This hope for a newly empowering, just, and equitable society sits at the core of this book's vision of our virtual future. I believe that the rise of so-called post-human technologies will soon produce robust virtual societies that will transform the way we live on Earth, while redefining what it means to be human. Picture a world in which you could master a new skill in a single afternoon, using advanced simulation technology that can pack a decade's worth of trial and error into a two-hour span. Or imagine participating in a massive festival that meaningfully involves every single user of the virtual world in which it exists: a universe-scale celebration in which thousands or even millions of people have the chance to be the center of attention. United by a spirit of play and a sense of mutual participation in the same transformative experience, the participants in this world will feel like truly integral parts of something bigger than themselves, on a scale that no terrestrial activity can hope to match.

The virtual worlds that people like me are working to build will be centered around these sorts of useful and fulfilling experiences. In them, users will be able to interact with their friends, meet new ones, learn valuable skills, have exciting adventures, and participate in civil society. These experiences will offer people the chances to explore new challenges, express their creativity, and consistently find satisfaction, social uplift, and joy. The advanced computing technologies that will power these worlds will be able to generate valuable experiences with speed and precision, like machines that are built to produce human fulfillment.

The rewards of virtual society won't be only psychological ones. Before long, people will earn money in virtual worlds by performing an array of jobs that will match and exceed real-world jobs in terms of salary, accessibility, and satisfaction. The inevitable expansion of economic opportunities within these other worlds will have a transformative effect on human society. In a decade or two, the locus of our culture, economy, and society will shift from a single world—the legacy "real world," you could say—to many worlds.

While these other worlds will immerse the senses, the fact that they will look and feel "real" won't be what makes them valuable. These worlds will be valuable because they will remake our lives by extending the context of society into new realms, allowing for the transfer of wealth, ideas, identity, and influence—the building blocks of human social relations—between our current reality and the digital ones that we create. The combination of these realities, and the transfer of value between them, will comprise the digital metaverse.

This book is your guide to virtual worlds and digital metaverses: why they are important, why they are necessary, and why they will change society for the better. In it, I hope to provide a working theory of how the metaverse will create value for both individuals and society. With this theory in hand, we can then look at the ways in which this value might be maximized. In the process, I hope to move beyond the business and technical contexts of the metaverse and into the human context. My goal in these pages is to present a comprehensive explanation of why ideating within virtual spaces matters so much to our past, present, and future. While I hope that investors and entrepreneurs will find this book useful, I have written it with many others in mind, too: scientists, regulators, historians, content creators, and everyday people looking to reconcile the hype they've heard about the metaverse with some sense of why it will matter to their own lives.

You can consider this book an attempt to offer a historically grounded and practical theory of metaverses: how to define them, how to measure their utility, and how to understand their interaction with existing ideas. What are the fundamental forces that drive humans to create these other worlds? How will they evolve as they take digital form? Why do they matter to individuals and to society? In the first half of this book, I'll address these questions and explain why the metaverse is more than just the future of the internet: It's the future of human experience. Though this book assumes and builds on the work of anthropologists and sociologists, it does not seek to replicate that work, only to demonstrate that the utility of other worlds is an established fact.

In the second half of this book, I will take a more microscopic view of the digital metaverses that will soon come to affect all of our lives. I'll attempt to establish a set of guiding principles for creating a metaverse that is equitable, useful, efficient, and fulfilling. I'll propose an ideal organizational model for building a valuable metaverse; examine how social, psychological, and economic value are related within virtual contexts; and offer some thoughts on ideal modes of oversight and regulation for the metaverse. My goal with this section is to establish the parameters for an optimally valuable metaverse, as well as the best ways to bring it to life.

My vision and predictions for the future are rooted in practical experience. As an entrepreneur and computer scientist who has spent much of the last decade building complex virtual worlds and the infrastructure for the metaverse, I have direct insight into the technical and organizational challenges that we face on the road to virtual society. Perhaps more important, I've long been immersed in the company of entrepreneurs, investors, and builders working toward the problem of creating the metaverse. This book represents the best synthesis of what I've learned from them over the past decade.

When I was growing up, digital games let me learn and experience things that were inaccessible in the real world. These games instilled in me a sense of wonder and exploration. My experience with them was, in fact, the opposite of the stereotypical image of a gamer who wants to withdraw from the world. In the games I played, I wanted to go somewhere, to do more, to be more, and to feel more fulfilled. Often, I would return from my gaming sessions feeling transformed. I strongly identified with the children in *The Chronicles of Narnia* by C. S. Lewis, who would go through the wardrobe into another world of adventures and return with fresh understandings and new perspectives—so much so that, as a kid, I personally investigated countless wardrobes for potential interdimensional portals. (I didn't find any.)

Now that I am lucky enough to build virtual worlds for a living, I am more convinced than ever that engagement within them can and will change people's lives for the better—if, that is, we take the time now, at this crucial juncture, to base our plans for the future in a clear understanding of the value that these worlds can create for individuals and society. This value is not limited to entertainment or escapism. One of the greatest surprises of my career has been the incredible importance of simulated virtual worlds to the future of military planning and strategy. My experience building virtual training environments for real-world militaries has convinced me that these simulated spaces will have immense value to countless other fields of human endeavor.

You might be skeptical that the metaverse will create any value whatsoever. After all, recent history is littered with examples of wild predictions around various innovations that pundits tend to lump together: virtual reality, augmented reality, artificial intelligence, cryptocurrency, and, yes, the metaverse. Society is now continuously anticipating, arguing, and betting on technology and products that do not yet fully exist. In the process, pundits and prognosticators tend to latch on to certain narratives of the future, regardless of whether they are accurate or optimal. Those narratives then become disproportionately important in determining which ideas and projects get funded, built, and used.

As a result, tech soothsayers are often correct about the general direction in which society is headed while also being wildly wrong about the specifics of the journey from here to there. Just think of the dot-com crash and the demise of so many companies that, while in the right field, had the wrong model of how value was created therein. The public is not well served when a new technology's loudest promoters cannot clearly explain its point or its purpose. Opaque and breathless narratives of the future tend to breed cynicism and resentment. This state is where we're at right now with the metaverse.

In the media and among entrepreneurs and investors, virtual worlds have been discussed and understood primarily through the prism of video games, commerce, and entertainment. Many are adamant that the metaverse is the next big thing, and yet are maddeningly vague when it comes to defining both what it is and why it will and should matter. Their visions for the metaverse often seem to be rooted in the immersive worlds of fiction: the *Matrix* films, for example, or books and stories by Neal Stephenson or William Gibson. When pressed for an exact definition of the metaverse, these people generally describe high-resolution embodied 3D environments with rich interactive possibilities—in which users might shop, play, meet, learn, and love.

The best that can be said for these frameworks is that they are woefully incomplete. Outdated or superficial visions of virtual worlds and the metaverse tend to focus on the what, but not the why; the thing, but not its purpose; the opportunities that will be made available to us in embodied digital spaces, but not the reasons why we would want to pursue them in the first place. These conceptual gaps present a problem if you are trying to build a metaverse business, or to regulate the space, or even to understand the changes that are about to happen. In the absence of meaningful discussion of the value that virtual worlds will create for individuals and society, the metaverse can start to seem intangible and superficial.

This broad, shallow model is inhibiting our ability to perceive and shape the future. We need better models for the future, ones that are conceptually expansive, rooted in visions of social value instead of just corporate profit. We must know why we're building what we're building, and why virtual worlds and the metaverse are worth the effort. And if we can understand the *why* of the digital metaverse—if we can clearly articulate its purpose and its potential—then we can grow that vision into a world of its own, one that will represent and serve humanity at its best.

In this book, I will focus on the *why* of the metaverse, in order to emphasize the purpose that it will serve in our lives and for society in general. The digital metaverse that I envision is one that will create untold social, psychological, and economic value for its users and for the wider world. It is a framework for a rich virtual society that will enhance, not supplant, our lives here on Earth. Like the development of writing, or the advent of the computer age, the dawn of the metaverse will be a grand pivot point in the history of humanity: a manifestation of the age-old human impulse to create cultural technologies that can usefully enhance and transform our lives and societies.

We almost always fail to understand that culture adapts to technology in nonlinear ways. If you had told most investors at the dawn of the internet that, in twenty years, people would be trading badly drawn JPEGs for millions of dollars while obsessively photographing every meal, or that a system like blockchain could be developed by entirely anonymous individuals, nobody would have believed you. Our transformative technologies tend to assume their own velocity and direction, which is why we must focus our thinking on why these technologies will matter to our lives rather than what forms they will take. In the absence of intelligent, capacious models and responsible, proactive thinking by key stakeholders within the spheres of investment, regulation, and infrastructure, the process of building the metaverse will be one of waste and unforced errors. It's important to avoid those pitfalls.

If you want to see around the next corner, this book is for you. It's a book that I wish I had been able to read back when I was first starting out in this business and was eager for something that might guide my own thinking and efforts. I hope that you're able to use the following chapters to guide your own thinking about the metaverse, and to help you understand its concrete human purpose. I believe that the digital metaverse will rank among the most important changes that humanity has ever experienced, because it is likely to act as a fork in history. The ability to live in many realities at once will be a fundamen-

VIRTUAL SOCIETY

Chapter 1

ANCIENT METAVERSES

I n present-day Turkey, amid the rocky plains of South-
eastern Anatolia, juts a monument, thirty meters in diam-
eter, of immense age. The T-shaped limestone structures
dotting this ancient hillside, some standing as tall as five and a
half meters, are set around enclosures and painstakingly carved
with depictions of animals. More than 240 of these structures
have been uncovered by archaeologists, who, as of this writing,
have excavated a tiny percentage of the full site, which is known
as Göbekli Tepe. Taken together or separately, the megaliths
serve today as a dispatch from the Neolithic: a barely scrutable
glimpse of our past that still offers a relevant lesson for our future.

When this monument was erected more than 10,000 years
ago, the place from which I write here in southwest England
had barely tasted freedom from continent-scale ice sheets.

Woolly mammoths still clung to existence, and agriculture had not yet been widely adopted. Yet in this alien world of near prehistory, at least 6,000 years before Stonehenge was built, primitive humans constructed a series of extraordinary stone megaliths and decorated them with elaborate carvings.

Why were these structures built? In the context of their era, there was no earthly reason for them to exist. It would not, therefore, be unreasonable to wonder if their construction wasn't actually motivated by earthly reasons—if it was, perhaps, compelled by belief in some other world.

To the best of our knowledge, the spiritual world or other reality implied by this monument (if, indeed, its purpose was religious) does not actually exist. Yet the representational universe contained within Göbekli Tepe—this early "virtual world"—involved the movement of enormous masses of stone over the course of a thousand years, surely at a nontrivial cost for its hunter-gatherer creators. A task of such magnitude is never undertaken on a whim, but especially not in the harsh environs of Neolithic Anatolia, a time and a place fundamentally unsuited for frivolous architectural digressions. To these nomadic builders and their society more than 10,000 years ago, the world invoked by these megaliths must have mattered as much as, if not more than, the physical world in which they lived.

The megaliths at Göbekli Tepe may seem to you like a product of a distant and alien past. But I believe they represent a fundamental human impulse, a power that we still manifest today. The first monuments erected by humans weren't carved out of stone so much as out of ingenuity. They were living ideas birthed into existence through collective agreement, imaginary forces imbued with the power of life and death, virtual worlds created through the force of a society's collective imagination.

For 10,000 years, humans have found ways to make the unreal real, just by willing it so.

We have always been a species of worldbuilders. Since humanity first emerged, we have used ingenious means to exist simultaneously in multiple realities: the animal reality of our earthly lives and the elevated reality of the worlds we create with our minds. We've been designing these other realities now for millennia, with tools no more advanced than our language and our imagination. While we sometimes build stone monuments to mark our belief in other worlds, these worlds exist separate and apart from the structures we raise to commemorate them. We speak our virtual worlds into existence, and we sustain them by the force of our collective belief.

From a cursory vantage, the dusty stones at Göbekli Tepe may seem like crude monuments from a forgotten, foreign world. But if you examine the carvings in detail, a universe of images and meaning floods your vision: scorpions and snarling beasts, geometric patterns, gesturing vultures and headless humans. Imagine how meaningful the mythologies held by these people must have been in order for them to craft such intricate works into stone—how tightly their belief system must have been interwoven into their everyday existence.

Just as the megaliths of Göbekli Tepe suggest a dynamic interplay between a virtual world and everyday existence, the virtual worlds that we'll be discussing throughout this book are far from static stories that are disconnected from our daily lives. They are worlds that our society treats as real, ones that can be the sources of actual wealth, power, and identity in the physical world. We build and inhabit these other worlds today for the same reasons our ancestors built them eons ago: to generate fulfillment and value, to materially improve our lives on Earth. Rather than marking the gates to these other realities with monuments of stone, these days we create digital gateways that conduct us from one world into another.

Though we might not always consciously realize that we are building these worlds, our skill at doing so affects and informs everything we do as a species. This book is in part about how this fundamental human talent for worldbuilding will shape our future, and how the coming age of virtual society represents not a new and foreign phenomenon, but the continuation of ancient traditions serving intrinsic human needs. But before we move toward the future, let's let our gaze linger a bit longer in the past, and closely examine what we mean by virtual worlds and worldbuilding.

WHEN WORDS BECOME WORLDS

Creating models of reality is an essential part of high-level thinking. In order to survive and operate effectively in the world, we must be able to simplify and experiment with outcomes as we plan or make decisions. In so doing, we create and evolve worlds of ideas that exist apart from and in conversation with the embodied world. This process is so fundamental to our language and our cognition that we rarely stop to consider its centrality to our day-to-day lives.

"The limits of my language mean the limits of my world," Ludwig Wittgenstein wrote in *Tractatus Logico-Philosophicus*. When we verbalize these worlds of ideas, we begin to create social models of them, ones that people other than ourselves can begin to access and use. In a meaningful sense, then, our words create our worlds. The common phrase "visualize a better world"—one in which world peace is the norm, for example— fundamentally means to visualize a virtual world, one in which that idealized outcome has already come true, and then to use the truth of that world to create equivalent value here on Earth.

This individual capacity for worldbuilding can precipitate the creation of rich, detailed, intergenerational worlds that can inspire us to great and sometimes terrible accomplishments: the birth of great religions, for example, many of which came to have tremendous utility to society while also serving as the source of continuing pain. We often construct these social realities to fill a definable need: to explain otherwise incomprehensible events, to justify actions, to add additional excitement to our lives, or simply to lend order to the chaos and danger of life. As people come to believe in these other worlds, their faith expands the worlds' parameters, and these realms can, effectively, come to life.

We do all of this work not just because we enjoy the act of building and believing in these worlds—although doing so can indeed be very enjoyable—but because they serve individual and social purposes without which our societies could not function. Society uses embodied worlds of culture and imagination to create common purpose and handle the emergent complexities of interpersonal dynamics; society uses these structures to regulate avarice and ambition and direct human energies toward noble purposes. A world with no shared culture, no structures within which to harness ingenuity and create shared experience, would be a brutal reality in which life would be reduced to modes of sustenance and survival.

"Myth is language," wrote the anthropologist Claude Lévi-Strauss, "functioning on an especially high level where meaning succeeds practically at 'taking off' from the linguistic ground on which it keeps rolling." These mythic worlds "take off" and become socially constructed realities: other worlds that rely on participants across the spectrum mutually agreeing to believe that they exist and that they matter. As such, in a meaningful sense, these virtual worlds are brought to life by this mutual agreement.

These other worlds aren't alternative realities into which we choose to escape: They are *more* reality. They are found spaces into which we can extend, evolve, and improve our social structures. Even today, these living other worlds and the events that happen there can enrich, expand, and affect our economy, our culture, and our daily lives. Think, for example, of the art and culture created as a result of societies' belief in other worlds. The ceiling of the Sistine Chapel is a priceless work in our world and also, in a sense, a gateway to the virtual world that inspired it, just as the Göbekli Tepe megaliths were effectively gateways to the world that inspired their creation.

Modern-day financial markets also count as other worlds that create value for our own. In these markets, fortunes and reputations are won and lost when societies en masse agree to ascribe great value and power to assets that often have no inherent worth beyond that which their stakeholders agree that they have. Professional sports also qualify as living, vibrant worlds of meaning. Look no further than the lengths to which superfans go to manifest their devotion to their chosen teams. Fans live and die with the fortunes of their hometown clubs, even though the outcomes of the games that they watch with such intensity will make no direct difference to their daily lives. Impactful athletic contests can become venues where we can temporarily modify reality in order to solve or sublimate social issues. When some under-resourced country defeats a wealthy nation in a World Cup match, for instance, the victory is felt profoundly by the smaller nation. It can serve to salve wounds, create pride, instill confidence and good feelings. These matches are mechanisms for meaningful social cohesion.

Such is the power of a constructed world of meaning: Outcomes in a world in which people have actively chosen to believe can ultimately matter more than outcomes in the world in which they just happened to be born. These virtual outcomes create value that can be transferred back to Earth. One clear sign of this transfer can be seen in the price of things that are not of tangible value, but are considered rare or special for other reasons. The Shroud of Turin, for instance, is considered priceless not because of the inherent worth of the fabric upon which its famous image appears, but because society deems it valuable. That value is an expression of belief in another world that has become substantially important to our own. It is a *virtual object* that, in its function, is not dissimilar to the works of digital art that so many people today find hard to understand: a discrete, irreplaceable token imbued with value localized within a constructed world of meaning.

It might seem like a stretch to compare the Shroud of Turin to a non-fungible token (NFT). But the differences between the relics of yesterday and the virtual objects of today and tomorrow—between virtual worlds accessed by the ritual imagination and those worlds accessed by Wi-Fi—aren't as stark as you might initially think.

Socially constructed realities, which I regard as a form of proto-metaverse, have existed on every inhabited continent, in every society, throughout recorded history. They're generally composed of a few distinct elements: a society or grouping of humans; another world or reality involving events, identity, rules, and things that are deemed to be in some way real; and an ongoing transfer of value between the two, which grows individual and group fulfillment, wealth, and meaning.

These worlds aren't just good stories. To their believers, they are actual places where cause creates effect, where things actually happen. Over time, these worlds take shape in the minds of the people who have called them into existence; over time, these virtual worlds can become so consequential and persistent that they can come to feel as real and as meaningful as our own. From them, value transfers into our world in the form of social structures and cohesion, notions of identity, transformative experiences, ritual observances, and so on. These worlds become living, useful ideas that demand and reward intellectual and emotional investment from their adherents.

History is littered with relevant examples. Think, for example, of the billions of people across continents and eras, up to and including our own, who have believed in nearby worlds populated by magical beings—genies, elves, witches, ghosts—and have structured their lives, and to an extent their societies, around their belief in these worlds. Think of ancient Romans reading entrails or taking the auspices in order to divine the will of the gods, and offering sacrifices and statues to honor and appease them. The Romans used to refuse to engage in war unless the proper rituals had been carried out. If they received improper auspices, they would pack up and go home until things changed. Even commerce would stop upon the receipt of bad signs.

In his book *Parallel Myths,* which examines the similarities in the myths ascribed to diverse civilizations across the globe and throughout human history, J. F. Bierlein noted that myth "is a constant among all human beings in all times. The patterns, stories, even details contained in myth are found everywhere and among everyone." The similarities between many facets of these modern and ancient metaverses—worlds often created roughly within the same time periods, by societies and groups that would have had no means of communicating with one another—suggest that the act of world creation is a primal human ability. In the 1970s, the anthropologist Charles Laughlin and the psychiatrist Eugene d'Aquili proposed that our brains evidenced a "cognitive imperative" for using myth and ritual to cohere groups, teach lessons, and resolve binary oppositions. More recently, d'Aquili and the neuroscientist Andrew Newberg used fMRI imaging to demonstrate that engaging with religious other worlds may indeed fulfill some basic neurological need.

In *Myth in Primitive Psychology*, the anthropologist Bronisław Malinowski wrote of "a special class of stories . . . [that] live not by idle interest, not as fictitious or even as true narrative, but are to the natives a statement of a primeval, greater and more relevant reality, by which the present life, fates, and activities of mankind are determined." Though Malinowski was referring specifically to the worldbuilding activities of a Melanesian tribe on the Trobriand Islands, the observation would be valid even for more modern cultures. The impulse to commit to the reality of worlds forged from our collective imaginations is an ancient and universal one, as is the instinct to make these virtual worlds matter to and create value for the real world. Consider, for example, how the lore of other worlds has long been used to teach people valuable and practical lessons: Don't drink from this particular well, don't go into the woods at night. The value created by engaging in other worlds doesn't stop at warning people away from wolves and tainted water. Worldbuilding is no less an important part of humanity's foundational skill set than fire, language, agriculture, or any other tool we consider elemental.

Virtual worlds are not and have never been just games, and imagining, creating, and role-playing within them is not just an entertainment activity. In the present day as in antiquity, they are and have been fundamental human accomplishments: units of cultural technology that create substantial intrinsic and ex-

lasting stone: eternal homes for the body to signify the eternality of the next world.

Today, the lasting legacy of ancient Egyptian civilization—the pyramids—is simultaneously a legacy of the virtual world that sustained and animated that civilization. And yet, from a strictly rational standpoint, one can hardly think of a less useful way for the ancients to have spent their time, resources, and energy. Yes, geopolitically speaking, building a bunch of massive, impractical pyramids out in the middle of the desert was a highly visible way for pharaohs to flex their military might. But every hour spent constructing the Great Pyramid of Giza was also an hour spent not inventing some other technology that, had the ancient Egyptians applied their ingenuity to it, may have materially improved their quality of life.

It is tempting to explain this bygone society's belief in an evanescent other world as merely a sacred or superstitious practice observed by a relatively primitive people; it is also tempting, from our standpoint in today's productivity-focused era, to view the pyramids as something of a beautiful waste, a project perhaps worth admiring but certainly not worth learning from. This lens is both limiting and artless. Production output is not the only or even the best measure of social value. The binary of *productive* or *unproductive* falls apart when contemplating the creation of an imagined reality in which "things" are not the primary output. If you insist on viewing the world exclusively from the standpoint of productive utility, not only will you fail to understand the past, you will fail in your efforts to create an optimal future.

The truth is that the ancient Egyptian fixation on the beyond can be understood outside of any religious or mythological context; it can, in fact, be seen as an entirely rational social choice. Not only can other worlds help to cohere and engage a society and its members, they can generate significant social, cultural, and economic value. It's not hard to grasp this point even today. The value that the pyramids continue to create as beautiful, soul-stirring cultural monuments is hard to dispute. It's inconceivable that anyone today would characterize them as a frivolous waste.

The metaverse will be a monumental cultural and techno-logical construction that will require enormous amounts of en-ergy, time, and attention, just like the pyramids required from the ancient Egyptians. One major argument against digital re-alities is that they may end up taking us away from the "real" world. But just like the pyramids, I believe that the metaverse will create enormous amounts of value to be transferred from the other world back to our own—value that will benefit society at large, not just the creators and arbiters of the other worlds.

In order to understand how digitally rendered virtual worlds will exceed their predecessors in this respect, we must first look at the various ways in which analog virtual worlds have served humans for millennia. You can probably think of some exam-ples off the top of your head. Virtual worlds have helped to ex-plain our own world, as in the various creation myths of the world's great religions. These worlds have been used, rightly or wrongly, to justify actions being taken by social leaders. They've been used to embed cultural values, teach moral lessons, or oth-erwise shape the contours of perception. As in the worship of sports teams, they can add a certain frisson to life, thus enliven-ing the doldrums of daily existence. A devoted fan will enhance her own experience of the sporting world by purchasing team apparel, or buying tickets to attend big games in person—an example of how virtual worlds can also create economic value for the societies that sustain them.

The fact that many of these worlds entertain their participants, while not their main social function, cannot and should not be discounted. Many virtual worlds originated as myths or stories, narratives devised in order to cohere or console a social unit during long, cold nights or some endless stretch of days. But stories are divided between teller and listener, author and reader, creative artists and passive recipients. A virtual world forged from human ingenuity erodes the difference between the two parties and turns everyone into an active participant in the life of the tale. A story in which all participants have the chance to be the protagonist isn't a story at all: It's a world.

The practice of constructing worlds so immersive or alluring that they come to feel real to their participants is better understood as a fundamental human impulse toward *ingenuity*: applied creativity or problem-solving. Once the parameters of a world have been created, then all new inputs to the world must work within that rule set. Additions to the world cannot blithely ignore existing information. No matter how creative the content gets, it still needs to play by the rules of its world in order to be compatible.

Unlike a myth or a story, a virtual world is an active process, not a fixed narrative. As such, it can "act back" and respond to the physical world. Things happen in the virtual world that feel real enough to its believers to affect goings-on in the real world. Meanwhile, believers and participants have the power to set and advance the boundaries of the virtual world, and to create meaning and order in the virtual world through their words and deeds. In his 2012 book *Legends of the Fire Spirits*, Robert Lebling wrote about the ongoing role of *jinn*, or genies, in Arabic culture and civilization, and by extension the ingenious ways in which those who believe in the jinn have established the parameters of their magical realm. "Jinn populate the world with offspring. They too are male and female and raise families," wrote Lebling. "They possess free will and make choices in life. They accept or reject God. Some become renegades and are classified as demons or ghouls. Others practice established religions and live what humans would consider conventional, 'normal' lives among their own kind." By imbuing jinn with a spectrum of human traits, believers in the jinn make the other world more approachable and create more points of entry for human engagement with that world.

Virtual worlds rooted in Western culture are no strangers to this sort of active, participatory worldbuilding. The Christian concept of heaven is a quintessential virtual world: an actual place, inaccessible by human technology, that believers hope to go to when they die. For centuries, the Roman Catholic Church has kept a sort of incomplete roster of heaven's inhabitants by canonizing certain deceased believers as saints, and thus verifying that those devout souls now live in heaven. Moreover, the Church has assigned certain saints to serve as patrons of specific causes on Earth, as anyone who has ever prayed to St. Anthony to help find a lost object can attest.

These acts of human ingenuity serve to expand the parameters and utility of a virtual world. In *Parallel Myths*, Bierlein quotes the psychologist and philosopher Pierre Janet's observation: "We are not to suppose that religion ever could have persisted if the gods had not spoken," meaning, in part, that other worlds are effectively useless to us if those worlds do not interact with our own. Assigning saints as patrons of human causes creates new opportunities for interactions between Earth and heaven, and thus serves to sustain both worlds. It actively acknowledges real events, thus changing real society; this is the inherent value transfer from heaven to here.

The saintly record-keeping described above also functions as a sort of crowdsourced historical record, a shared set of events and dramatis personae upon which all participants in the world can agree. A virtual world matters more to its adherents when it is a persistent world that outlives its creators—when it boasts a substantive history on which everyone can reach a consensus. The outcome of the World Cup, for instance, matters in part because of all the other World Cup winners; today's outcome exists in line with yesterday's outcomes, and this continuity of consequence creates depth of belief. By creating an open, accessible log of what happened when, this process also builds trust in the fidelity of the world and the information therein. In that way, come to think of it, it's not entirely dissimilar to the growing world of decentralized ledgers, or blockchains, which allow groups of participants to agree on a sequence of events or an accounting of ownership. (More on blockchains and the metaverse in Chapters 7 and 8.)

By now, we've established that people expand the boundaries of their virtual worlds via acts of ingenuity. But why does this matter, and what purpose is served by these ingenious acts? Why does it matter to society that some jinn wear hats while others go bareheaded? Why is this intricate worldbuilding good for both the world and its builders?

Like agriculture, or constructing shelter, worldbuilding is a fundamental human task that people undertake in order to survive and thrive. When a story becomes a world to explore, contour, and enliven by the force of your belief, then it can transcend mere entertainment value and begin to serve a psychological function for its participants. Active participation in the life of a virtual world requires the sort of applied creativity that engenders feelings of intrinsic fulfillment.

In this regard, socially constructed realities are cultural technologies created by humans to help enhance and order their lives. They represent another stage on which society conducts its business. Wins and losses in the world of sport impact relations between countries, and can act as another form of diplomacy. Imbuing certain people with fame creates models to follow for others. We use the "other world" to magnify the psychological value to us of our everyday lives by infusing ordinary things and occurrences with meaning. We "play in a virtual world" and score "virtual points" with social meaning because something deep inside us compels us to do so. Fulfilling these needs correlates to positive mental and emotional outcomes. These sorts of positive outcomes were, in part, the point of the pyramids. By investing in the reality of their ancient metaverse, the Egyptians found ways to directly improve the substance of their lives on Earth.

Scholars of myth have long argued that myth gives structure and meaning to human life; that meaning is amplified when a myth evolves into a world. A virtual world's ability to fulfill needs grows when lots and lots of people believe in the world. Conversely, a virtual world cannot be long sustained by a mere handful of adherents. Consider the difference between a global sport and a game I invent with my nine friends and play regularly. My game might be a great game, one that is completely immersive, one that consumes all of my cohorts' time and attention. If its reach is limited to the ten of us, though, then it's ultimately just a weird hobby, and it has limited social function. For a virtual world to provide lasting, wide-ranging value, its participants must be a large enough group to be considered a society. When that threshold is reached, psychological value can transmute into wide-ranging social value.

The sociologist Émile Durkheim, in his work on the concept of collective effervescence—effectively, the sense of emotional uplift shared within a group of people all involved in the same ritual action—divided daily life into the sacred and the profane. The profane part of life comprises our day-to-day routines, the people and objects with whom and which we routinely interact, all of the tasks and jobs and responsibilities that consume our waking hours and offer little in the way of transformative excitement or joy. In other words, the profane is the mundane. The sacred is the opposite, kept apart and often forbidden. When large groups of people come together to access the sacred in the course of celebration, commemoration, or ritual observance, the group commitment to and belief in the meaning and relevance of the event can create collective effervescence: uplifting, changing, and binding together all those who participate. We need only check Twitter during the Super Bowl or the World Cup finals to observe a community cohering out of collective attention on a meaningful event within a socially constructed reality.

The ritual function of socially constructed realities also has a practical function in the lives of the societies that create them. These social realities, historically, are accessed via ritual, and participation in the ritual creates *communitas,* in the term of anthropologist Victor Turner: a sense of humanity and relatedness that is linked to the hierarchical structures that govern our daily lives. What happens in a virtual world when those structures are upended? In the foreword to *The Ritual Process: Structure and Anti-Structure,* written after two and a half years spent observing the rituals of the Ndembu people in Zambia, Turner observed that "in order to live, to breathe, and to generate novelty, human beings have had to create . . . liminal areas of time and space—rituals, carnivals, dramas, and latterly films—[that] are open to the play of thought, feelings and will; in them are generated new models, often fantastic, some of which may have sufficient power and plausibility to replace eventually the force-backed political and jural models that control the centers of a society's ongoing life."

When, for example, we agree to believe that genies are real, we create an "anti-structure" that offers an alternative ordering of society and reality, one which presents different hierarchies and priorities than those that normally govern our lives. This anti-structure rewards modes of thought and perception that differ from those we deploy in the real world: nonlinear, dream-logical, playful, transgressive methods. This phase change from structure to anti-structure frees the participants in the ritual to think, perceive, and invent differently. They can then transfer those outputs back to the real world.

Göbekli Tepe is the earliest apparent example of this sort of phase change in action. While scholars do not know exactly why the megalithic structures there were built, they have suggested that the monuments may indeed have been religious in nature. But just as with the pyramids, the value created by our predecessors' commitment to this parallel reality can be observed and understood outside the context of religion or superstition. In a 2008 *Smithsonian* article, archaeologists excavating the megalithic site postulated that the sustained effort and coordination required to build Göbekli Tepe roughly 12,000 years ago may have precipitated the contemporaneous "Neolithic Revolution," in which the era's hunter-gatherers transitioned to living in fixed settlements and developed complex societies. In other words, while the megaliths may have begun as an effort to commemorate or signal to an undefined spiritual realm, they brought about a new era of civilization right here on Earth.

WHEN A WORLD BECOMES A METAVERSE

The apparently civilizing effect that Göbekli Tepe had on the Neolithic exemplifies the primary and most important social function of a virtual world: the way in which value transfers and transmutes from that world to this one. Here, too, it is useful to refer to ritual practice as a way of illustrating the point. Many social rituals reach apotheosis when participants consume the flesh of an animal or some other sacrifice. The consumption of the flesh holds value beyond its caloric content—i.e., it differs from dinner—because of the meaning that has been assigned to the flesh by all subscribers to the other world. The act of consuming the flesh, then, creates a bridge between the two worlds, across which meaning and value are transferred from one to the other.

Since a virtual world is not merely a story, the living ideas therein are granted a measure of agency by their creators. The gods and the jinn have their own priorities, so to speak, and those priorities do not always align with our own. In Iran, Robert Lebling wrote, "fear of revenge by the jinn makes many people careful. . . . When a child suddenly begins crying or acts frightened without reason, many Iranians conclude that he must have hurt a jinn baby and that its mother is retaliating. If the child's mother is present, she must express some of her breast milk on the spot where the child was sitting. This generosity will please the jinn parent and she will stop punishing the human child." As a world and its adherents both grow, the living ideas therein act back on the society that created them, and the act of creating the world evolves into the act of maintaining a conversation with that world.

Over the course of that ongoing conversation, value is transferred from that world to our own, often beginning with ritual observance or experience and expanding out from there. At Göbekli Tepe, as we just read, the conversation between worlds initially inspired the construction of megaliths with a presumed ritual function. In order to build the structures, though, the nomads had to settle down. They had to construct villages, which in turn led to new forms of social organization.

Virtual worlds change the real world in ways that we cannot really predict when we create them. All that we can reasonably predict, in fact, is that those other worlds *will* in fact change our own given sufficient time and depth of belief. You could even say that we create these worlds precisely so that they will change our own. If over the course of history we have done this subconsciously, in our digital future we will do so with intent, creating virtual worlds rich with experiences and meaning in order to put them in conversation with our own world and realize value from the ensuing changes. This bilateral value exchange among the worlds in a set is what makes those worlds a metaverse.

My most basic definition of a metaverse is a conversation, a structure of multiple worlds that permits value exchange between them. (I will offer a more thorough definition of the metaverse in Chapter 5.) On an elemental level, this exchange requires neither advanced technology nor digital simulations in order to happen. Like the ritual consumption of flesh, the value transfer is powered by community consent and acceptance of the value of the other world. This belief allows the other world to catalyze change in our own.

Belief in magic, miracles, and spectral creatures is a persistent theme of human history, and a persistent example of socially constructed realities, in which widespread belief in the importance and existence of a thing effectively serves to make it real. In Iceland and the Faroe Islands, many residents have long professed belief in the Huldufólk, or "hidden people": human-oid elves who live in a parallel world that interacts with our own. The Huldufólk, visible only when they wish to be seen, are said to have the power to confer fertility or famine; they are also said to be fond of throwing raucous parties during the Christmas season. Their world is directly conversant with ours.

Far from being a relic of a simpler, more ignorant era in Icelandic history, the world of the Huldufólk is still relevant to the Iceland of today. Surveys consistently show that a not insignificant percentage of Icelanders are at least willing to acknowledge that the Huldufólk may exist, and some journalists believe that these surveys routinely underestimate the number of residents there who remain invested in the reality of the hidden people. In 2013, for instance, construction of a road from a Reykjavík suburb to the Álftanes peninsula was halted when a group of elf advocates sued to stop the road from being built, claiming that it would cut right through the elves' habitat. Writing on the case, the *Independent* observed that the Huldufólk "affect construction plans so regularly that the road and coastal administration has come up with a stock media response for elf inquiries."

The world of the Huldufólk is a socially constructed reality that, even today, matters enough to regularly affect daily life in Iceland. It's real enough to stop a road from being built, and while skeptics might consider belief in the Huldufólk to be an unnecessary impediment to progress, the proposed road mentioned above would have also cut through many animal habitats. In a 2020 essay in the *Georgetown Journal of International Affairs,* Bryndís Björgvinsdóttir noted that, in modern Iceland, "elf belief is almost exclusively expressed in terms of construction projects where nature is being transformed by humans.

The sacrosanctity of natural sites spurs conversations about the value of nature, our environment, and the future of humans in nature." The mirror world helps catalyze conservationists' efforts in the real world.

The value transfer from one world to another often takes that form, where the other world offers a necessary pretext for taking desirable action in this world. Here's another example. It was 585 B.C. and, as usual, the Medes were at war with the Lydians. The two neighboring kingdoms had been fighting for six long years, and the conflict had long since become a war of attrition. As Herodotus wrote in his *Histories,* "the Medes gained many victories over the Lydians, and the Lydians also gained many victories over the Medes." To its participants, it must have seemed at times like the war could be stopped only by divine intervention.

That's just what happened as the warring factions were facing off in what is now Turkey. As Herodotus told it, the battle had just begun to heat up when, all at once, a dark shadow crept over the sun. Before long, the battlefield had been plunged into darkness. It was a solar eclipse—but, to the warring armies, the sudden transition from day to night was nothing less than a divine portent, a sign of the gods' displeasure. All at once the Medes and the Lydians "ceased fighting," wrote Herodotus, "and were alike anxious to have terms of peace agreed on." Those terms were quickly established. The Halys River was thenceforth established as the border between Media and Lydia; the peace was consummated when the Lydian king agreed to betroth his daughter to the son of the Median king. And they all lived happily ever after, or at least until the Medes were conquered by the Persians thirty-five years later.

Consider the conditions necessary here for the relevant parties to reach this happy outcome. The Medes and the Lydians, two warring societies, first had to both agree on the reality of another consequential realm that existed in conversation with their own. Then they had to agree that the denizens of that other realm took a keen interest in the geopolitical turmoil of Earth. Then they had to agree that a decision made in that other realm could have tangible effects on Earth. Then they had to agree to change their plans and behavior on Earth as a means of appeasing actors in that other realm.

The results of meeting these conditions were numerous and positive. A battle ended abruptly. A six-year war concluded. Countless lives were saved and a marriage began, all because the Medes and the Lydians had agreed to believe that the gods had turned day into night to communicate their displeasure. Pragmatically speaking, it is very possible that, after six years of relatively fruitless fighting, the two warring parties were simply tired of combat and ready to seize on any possible excuse to bring the conflict to a close. But they clearly needed a pretext in order to achieve that outcome. And without mutual investment in this constructed reality, the eclipse wouldn't work even as a pretext.

The outcome of the Battle of the Eclipse, as it later came to be known, wasn't just an accidental byproduct of people believing that an eclipse was a portent from the gods. It's a product of an ancient metaverse in which actions in the physical world were thought to prompt reciprocal actions in the virtual world; those actions in turn caused people in the physical world to change their behavior. It's a function of two worlds in constant conversation with each other—and it's anyone's guess as to which of those two worlds is the more potent.

THE LIMITATIONS OF OLYMPUS

In the introduction to this book, I argued that we will soon be able to build digital worlds so present and immersive that they may be broadly indistinguishable from reality—not just in terms of how they look and feel, but in terms of how much they matter to the people who use them. Not only does this metaversal future bring with it the possibility of creating more fulfillment, meaning, and value for more people and societies than ever before, it will also allow us to construct other worlds that are more transparent and democratic than their predecessors.

While I have focused in this chapter on the positive social effects of the virtual worlds and ancient metaverses that humans have built and tended over the years, not every virtual world is a good one, and not every act of metaversal value transfer creates positive change. The dark, or at least checkered, metaverses of world history have sometimes brought fear and repression to the world, and have been used by the unscrupulous to obtain and hold power over others, while exploiting their own positions for personal gain. These dark worlds are almost uniformly characterized by a lack of transparency regarding rules and priorities, opaque and shifting histories, and tight control over ritual function and participation. It's not that, per Pierre Janet, the gods of these worlds stop speaking to mortals—it's that only a small number of people are empowered to hear and interpret their voices.

Let's consider the seventeenth-century witchcraft scare in Salem, Massachusetts, within this framework. Today, most of us would agree that necromancy does not actually exist in our real world, and that the spate of trials and executions designed to eliminate witches from colonial Massachusetts in 1692 and

1693 is best understood as an example of mass hysteria, misogyny, and, perhaps, widespread ergot poisoning.

At the time, though, it would have been reasonable for your average Salemite to believe that witchcraft was real and that witches walked among them. Everyone else in Salem seemed to believe it, after all: judges, preachers, reasonable and respectable people. The individuals put on trial for witchcraft, innocent though they surely were, could not close their eyes and escape their travails. An entire city chose to believe that witchcraft was real, and that many of their friends and neighbors were in frequent communication with another world, one that bestowed upon them dark powers to be used on Earth. For two unfortunate years, the citizens of Salem lived and believed in this unfortunate other reality. They could not wish the witches away, so they instead decided to try them and kill them. And for something to die, it first has to live.

The reason why Salem has become so notorious isn't just because the residents there let their belief in witches lead them into social hysteria for a few years in the late 1600s. It's also because the religious and juridical authorities of Salem became the arbiters of who was and who wasn't a witch. When interactions with these other worlds are mediated by a chosen few, a metaverse can soon go sour. Consider Europe before the development of movable type, when religious texts were mostly in Latin and access was tightly controlled by the clergy. In this era, very few people could read the "heavenly ledger" detailing the workings and priorities of the other world. Those few people who *could* read it held immense power.

Though the eclipse that halted fighting on the Halys was the sort of portent that all could see and interpret, other portents in antiquity were not always quite so accessible. A caste of priests controlled access to Olympus, and to the wishes and whims of its godly residents, through haruspicy and other divination techniques that outsiders could neither parse, penetrate, nor challenge. A small group of people would assert that something had happened in a world that existed only by virtue of collective faith in its existence, and a larger group of people would go along with that assertion. In the process, everyone was sort of making things up as they went along.

The main limitation of Olympus, then, is that for most people there is a cap on the value that can be created within and transferred out of an essentially passive virtual world. The scope of activities and distribution of meaning is undemocratic. There is a strict barrier between you and the gods, and while the gods can, on occasion, come to visit you, you cannot go to visit them. You cannot explore Olympus; you cannot discover its truths for yourself. You, as a regular person, are reliant on what other people tell you about Olympus and its inhabitants. You are disempowered, in other words, and a disempowering world is a fundamentally dysfunctional world.

The modern metaverse—the network of virtual worlds conversant with our own that will comprise this book's primary subject matter—can improve on its ancient predecessors while still filling the same innate human needs. The digital worlds of the modern metaverse are and will be consequential, instantiated, long-lasting places that can support large groups of participants. They will fill meaningful psychological needs for their participants, and they will act back on the outside world as they have done throughout human history, thus creating and transferring value from one world to another.

There will be many differences between a modern virtual world and these ancient metaverses. The worlds we create with computer code will be infinitely more complex and immersive than their predecessors. They will be more capacious and accessible than ever. They will be places that you can visit, places where the parameters will be clear and understood by all. You won't be reliant on a high priest to tell you what's happened and why. These worlds will have understandable and explorable rules, which means that you will be able to find your own valuable truths therein.

This futuristic metaverse, so richly defined and dense with meaning and opportunity, exists on a continuum with the metaversal past. But the virtual worlds of the future will be Olympuses that you can climb. You'll be able to explore the terrain, test the limits of your abilities, commune with the gods, and, in a sense, even become one yourself. Humans have worked to create and invest in these sorts of other worlds since they first began to gather in groups. In the very near future, for the first time ever, we'll all have the opportunity to take starring roles, to make these worlds our own.

At first, it may feel strange to opt into these new possibilities. Modern society, especially in the West, has for more than a century now worked to elevate the profane over the sacred, and we are all products of the societies in which we were raised. Göbekli Tepe, ancient Egypt, the Roman Empire: These historical moments seem foreign to us in part because our society is not currently organized around constant communication with gods and monsters. Instead, it is organized around the clock, and modes of productivity and self-improvement. Though our physical output on Earth has increased under this regime, our reliance and acceptance of the merit and power of constructed worlds of meaning has rarely been at a lower ebb.

If anti-structure exists in the West these days, aside from organized religion and valuable social fictions such as professional sports and the stock markets, it exists in the form of digital games. These spaces allow their participants to invert the hier-archies that govern their daily lives and access and create differ-ent forms of meaning than are readily available at home or on the job. The immersive digital games of today are the building blocks of the fulfilling, valuable digital worlds of tomorrow, and it's important that we understand both how they function within a historical tradition of constructed and ingenious worlds of meaning, and the important roles they fill within the context of a society that has come to disparage and reject the value of worldbuilding, play, and free time. In order to under-stand why the rise of virtual society will be a good thing, you must understand how and why today's production-focused so-ciety wants you to think that it won't.

Chapter 2

WORK, PLAY, AND THE PURPOSE OF FREE TIME

"I don't want to work / I want to bang on the drum all day." With this classic couplet from his 1983 song "Bang the Drum All Day," Todd Rundgren captured the complicated relationship among work, leisure, and personal fulfillment in modern society. Many of us spend our days working at jobs that we don't particularly enjoy, while counting down the hours until we can pursue the leisure activities that give our lives meaning. But we don't often stop and think about why our lives are structured like this, with productive labor front and center, and individual fulfillment consigned to the margins. Why isn't it the other way around? Why *can't* we bang on the drum all day?

In previous eras, as I wrote in the last chapter, many societies lived in tandem with mirror worlds of their own design, worlds often accessed by rituals that inverted the hierarchies of daily life. Today, due to the unique cultural circumstances of our hyper-industrialized era, society is firmly centered around principles of productivity. Save a handful of outlier cultures, all people from the very rich to the very poor are incentivized to maximize their own productivity, and truly "free" time is scarce. In previous eras, though, the wealthy lived lives of leisure and the working classes aspired to do the same. "Leisure is essential to civilization," wrote Bertrand Russell in his 1935 essay "In Praise of Idleness," arguing the merits of seeing labor as a means to an end, rather than as an end in itself.

The parameters of the optimal "work-life balance" have evolved over time, in tandem with the changing nature of the economy. Western society no longer offers the abundant opportunities for *communitas* that, throughout history, have counterbalanced the demands, pressures, and hierarchies of the workaday world. As we stand at the brink of the coming age of virtual society, and all the social and economic changes that this new era will bring, we have a unique opportunity to reexamine these parameters—and to redefine the nature of both work and play while optimizing for peak fulfillment rather than just peak productivity.

In this chapter, we will examine why we spend our time the way we do, and I will identify some of the ways virtual worlds will help make that time more fulfilling, more interesting, and, ultimately, more productive. I'm going to tell you about our current best understanding of the psychological underpinnings of human motivation, and why games and virtual worlds are built to confer motivation in ways that the outside world just can't match.

Lots of people have argued that games are bad for you—and, indeed, some games *are* bad for you. The best virtual worlds, however, are gymnasiums for the mind, environments that are purpose-built for human fulfillment. An hour spent in a high-quality virtual world shouldn't be thought of as time wasted; rather, as I'm going to argue, it's reasonable to see that hour as an exercise in self-improvement.

Your mind needs nutrition just like your body does. It needs to be challenged, encouraged, and fed with problems to solve and skills to develop. For many of us, though, our jobs offer a steady diet of empty calories. More and more people are getting their requisite brain food during their free time, from gaming environments and virtual worlds where the process is the product. By the end of this chapter, I hope to have challenged some of the assumptions you might have about what counts as a "healthy" use of one's time in modern society. But, before we can get there, it's important to understand why it feels like we are working so hard all the time these days—and how both labor and leisure have evolved over the course of the past two hundred years.

THE DECLINE AND FALL OF THE LEISURE CLASS

From the dawn of the Industrial Revolution through the Gilded Age—a time of great industrial productivity, fueled by the efforts of countless menial laborers working long hours for low pay in unregulated environments under often dangerous conditions—conspicuous idleness was a status symbol for the very rich. An entire generation of Western literature concerns the follies and foibles of the gentry: the Emma Woodhouses and Mr. Knightleys who already had all the money they could possibly need, which they spent on an endless succession of balls, dinner parties, and marriage schemes. At the turn of the twentieth century, American industrialists and their families and hangers-on flocked to ornate mansions in posh vacation en-claves and spent their time cultivating talents for riding, sailing, music, dancing, and drinking profusely at lunch.

Blue-collar leisure in the industrialized world was viewed differently. There was less of it, for one thing, and it was meant to be primarily devotional or productive. The most consistent leisure opportunities came on Sundays, when, as per Christian tradition, rich and poor alike were expected to spend their time in worship, rest, and contemplation. As the twentieth century progressed, workers were urged by do-gooders and the media to adopt hobbies, such as collecting or model-making, so that their spare time would be filled with self-directed, process-oriented labor, instead of with what moralists assumed would just be idle drunkenness. (Have you ever wondered why our weekends are two days long instead of just one day? While the labor movement can take most of the credit for persuading em-ployers to provide two full days of rest, one intriguing historical thesis posits that the two-day weekend was also instituted to stop workers from getting drunk on their lone day off—Sunday—and then skipping work the next day to nurse their hangovers. If they had Saturday to use as their drinking day, those work-ers could sleep it off on Sunday and be at work when the whistle blew on Monday morning.)

The American anti-saloon leagues and temperance move-ments of the era were primarily focused on inhibiting blue-collar dissolution; it was sort of expected that the very rich would spend their time lost in a frivolous champagne haze. The poor, meanwhile, were worked half to death, and given occa-sional holiday respites as a reward for and reinforcement of their constant hard work. The laborer's dream was to get rich enough to be able to do absolutely nothing of consequence with their time.

Things have changed. As Yale law professor Daniel Markovits argued in his 2019 book *The Meritocracy Trap*, a hyper-productive and hyper-efficient ethos dominates the mindset of today's rich and powerful. Among other things, these habits can reassure them that their wealth is a product of diligence rather than family background and social advantages. Many modern CEOs cultivate habits of sleep deprivation as they sit in meetings from morning to midnight while occasionally running multiple companies at once. Tesla CEO Elon Musk, incredibly, once claimed to work 120-hour weeks, thus putting to shame the countless strivers who undertake a mere 100-hour work-week. (Despite being a CEO of a startup myself—prone, unfortunately, to long workdays—I have yet to comprehend how Musk was counting those 120 hours. By my estimation, such a schedule would not only render leisure time entirely impossible, it would also likely require defecation at one's desk.) Until the practice was ruled illegal by the Supreme People's Court in October 2021, many Chinese companies expected their workers to abide by the "996" work culture: 9 A.M. to 9 P.M., six days per week. Financiers and investors spend long hours developing and tweaking complex algorithms to eke out small edges on the financial markets. Attorneys, consultants, and other professionals often advance in their careers based on the number of billable hours they can generate. Meritocratic society no longer holds Sundays sacred, and the day is no longer seen as one of enforced rest.

This grind-and-hustle ethos trickles down from the professions to the rest of the economy. Since so many businesspeople and professionals are working so much, lots of other service-based businesses also must be open all the time, in order to be there when their customers need them. Where their forebears seemed to delight in working as little as possible, for today's rich and powerful, life is an endless quest to optimize and extend their own productivity.

In the case of blue-collar jobs, while labor laws, social reforms, and technological improvements have made many workplaces less dangerous, a suite of economic, political, and demographic changes in many cases have also made working-class employment less stable, lucrative, and fulfilling than in decades past. White-collar workers, too, no longer live in a world of three-martini lunches and fat expense accounts. Automation, computerization, and outsourcing; a general corporate emphasis on maximizing shareholder value; widespread wage stagnation; and the ongoing shift from an industrial economy to a data economy have transformed the nature of both blue- and white-collar middle-class labor—as well as middle-class leisure.

Today, while the number of hours worked by the average middle-class worker has dipped over time—a 2017 study from the Organisation for Economic Co-operation and Development showed that the average hours worked in seven fully developed countries declined from 1970 to 2017 (though that same metric has started to rise again recently in some countries)—the quality of those hours, in terms of individual fulfillment, has also substantially declined. The late anthropologist David Graeber argued that automation has created a vast reservoir of "bullshit jobs," in which many people "spend their entire working lives performing tasks they secretly believe do not really need to be performed." The trajectory I've just described—with both managers and laborers urged toward effectively perpetual toil, accompanied by a decline in idle leisure on the top end and a decline in the quality of the hours worked by those lower on the corporate org chart—is the natural and inevitable evolution of the ways in which, centuries ago, the industrialized West came to orient its time around modes of productivity.

Time once moved differently, particularly in agrarian societies. Once the harvest was sown, throwing an additional thousand hours of labor at the fields wouldn't help the crops to grow any faster. It's no coincidence that the pyramids and other great public works projects in ancient Egypt were built against this unused time, in the yearly cadence of farming and the flooding of the Nile. Of course there's more time to commune with the mirror world when it's the "off season" in your own.

The advent of the Industrial Age changed that relationship among time, labor, and capital. Factories could produce around the clock, and they could do so with greater speed and volume than ever before. A machine that runs twelve hours a day will produce more widgets than one that runs for only eight hours per day—and a machine that runs twenty-four hours per day will produce the most widgets of all. As such, at many factories, the workday is divided into eight-hour shifts, so that there will always be people on hand to keep the widget machines humming. Industrialization raised the potential value of every single work hour—the more hours you worked, the more widgets you produced, and the more money you made—and thus wages became tied to effort and production. Labor, previously guided by harvest cycles, became clock-oriented, and society started to reorganize around new principles of productivity.

This shift wasn't all bad. Increases in industrial productivity created prosperity, extended life spans, and raised the standard of living throughout industrialized economies. Ordinary people benefitted from having a lot of newer and better things to buy, and a lot more employment opportunities than were quantifiably available in the era of pre-industrial peasantry. In exchange for these gains, though, many people sacrificed autonomy, which they had known as artisan and agrarian laborers, to control the contours of their workdays. As societies industrialized, they tacitly urged their citizens to view themselves as constituent parts of a grand machine, one that produced things for everyone's collective betterment.

The bargain was a good one until the terms suddenly changed. Part of the point of industrialization, at least as understood by utopian theorists, was that automation would reduce individualized labor burdens and create wealth and free time for average people. "In a world where no one is compelled to work more than four hours a day every person possessed of scientific curiosity will be able to indulge it, and every painter will be able to paint without starving, however excellent his pictures may be," wrote Bertrand Russell, arguing that society should use industrial technologies to shorten the workday and make all men and women members of the leisure class. What happened in practice was very different. As the grand machine grew more and more efficient, producing more and more things to higher standards in less time—and, just as important, as other parts of the world caught up to the West and began to compete in world markets—the pursuit of productivity was decoupled from the pursuit of a better, richer life for workers.

According to a 2015 study from the Economic Policy Institute, worker productivity and hourly compensation in the United States rose at a virtually identical rate from 1948 to 1973. But from 1973 to 2015, hourly compensation rates stagnated while productivity continued to skyrocket—meaning, effectively, that individual and societal prosperity were no longer on parallel tracks.

For many developed nations, productivity has become an end in itself. It is an end that primarily serves the few plutocrats who sit at the top of the economy; an end that exists apart from any ambitions of securing individual dignity and prosperity for those who serve as the cogs in the machine. The notion of productivity as an unreconstructed good is today inescapable. From the moment we begin our schooling, many of us are exposed to a relentless social narrative of productivity improvement, efficiency, economic output, and collective progress through creation. We live and work in societies that literally depend on endless productivity growth in their economic and political foundations; as such, we borrow against a future in which we hope we will be more productive than today. The pension funds that pay for our aging population are invested in and require that same promise of future growth. To be more productive, or inventive in ways that will make you more productive, is the nature and purpose of our most important work.

Though much lip service is paid by big companies to the concept of work-life balance, many corporate workers today are tacitly or explicitly encouraged to treat their work *as* their life. A century ago, business leaders urged their employees to adopt hobbies during their off-hours. Today, while CEOs might still encourage their underlings to step away from their desks in order to, say, play table tennis or take a yoga class, more and more these activities take place within the ecosystem of work. There's a ping-pong table in the office break room. The yoga teacher is on salary. Leisure time has become bait, luring workers into staying later at the office and working harder while they're there. Not only does it seem quaint to imagine modern CEOs advising their workers to go fly model airplanes on their days off, it seems quaint to imagine tech workers *having* any true days off.

The production imperatives of the modern economy have, in turn, created a deep skepticism with the worlds of social ritual, leisure, and play. It can sometimes feel like society actively dissuades people from pastimes with no measurable output, pastimes that you pursue not because of what you might win, learn, or gain, but simply because they are fun and make you feel good. But even if society no longer officially condones leisure, there's clearly still a desire for the release provided by social ritual and play. We can sense it in the fandoms surrounding works of popular culture, in which enthusiasts come together to discuss these works and extend their mythologies via fan-created stories and works of art. We can feel it in the meme-ification of discourse online; in the transgressive, anti-structural nature of conversing about important matters not via logical argument, but with a cheeky saying slapped over a photo of a cat. But the biggest indicator of this ongoing desire for play, to my mind, is the popularity of video games, which, as both an industry and a pastime, are a major aberration in the work-till-you-drop zeitgeist.

Gaming is a massive, multibillion-dollar industry, with worldwide revenues in 2020 sitting at just around $180 billion, according to MarketWatch—more than was earned by the global film industry and all North American professional sports combined. The aforementioned industries are differentiated by more than just their bottom lines. Watching a movie, for example, is a passive and finite activity. You can watch an entire film in the hours between dinner and bedtime. You can zone out while you're watching it, too, because it requires no viewer engagement in order to advance the plot. All you can do with a movie is watch it, and once it's over, it's over. If you watch it again, which you probably won't, it'll be the exact same movie as it was before.

Video games are different. The best games are active, expansive environments that are built to intellectually engage their users for weeks, months, even years at a time. For their most passionate devotees, playing a game can consume as much time as a job. Many gamers will come home from their job and put in the equivalent of another full day's work at their gaming consoles: not just passively consuming an entertainment product, but using their minds and reflexes to actively engage with the world on the screen. "I'm a fifty-year-old grandmother of five and an award-winning journalist with a respectable job at a local newspaper," a gamer named Alyssa Schnugg wrote for *PC Gamer* in 2018, in an article recounting her labor-intensive tenure as an *Ultima Online* Guild Master. "Every night I sit at my computer, boot up the classic version of *Ultima Online,* and my second job begins."

The sheer variety of games to which players become attached implies that games can offer more than just short-term dopamine hits. Some of my favorite examples of games that require an active, long-term commitment from their players are trucking simulators, such as *American Truck Simulator,* that mimic the environment you'd find behind the wheel of a big rig. Though these trucking simulators are technologically advanced as far as the graphics and driving dynamics go, there's nothing particularly complicated about their premise. Playing the game is like being in a truck, hauling goods down a highway, and it can take as much time to drive a route on the simulator as it does in real life.

Curiously enough, these simulators are popular among the ranks of actual professional truckers. That's right: A trucker might come home after spending a week hauling a load of cabbages across the country, then sit down at his simulator and relax by . . . virtually hauling a load of cabbages across the country. From one perspective, this trend could seem sort of depressing: a sign that we're so trained to value diligence that, even in our spare time, we can unwind only by mimicking productive labor. But these simulators often serve to remind truckers what they liked about trucking in the first place. Speaking to a reporter for *The Face* in 2019, one former trucker explained his fondness for *Euro Truck Simulator 2*: "You are your own boss, you have the freedom to choose any truck, cargo or destination you want. You can make it easy—or hard." Another trucker revealed that he enjoyed playing *American Truck Simulator* in order to experience driving different sorts of trucks and carrying loads that he'd never be able to haul in real life. It's work as play as work.

Why are these sorts of labor-intensive games gaining such traction in a world where many people are not only working longer and harder than ever, but are also subtly dissuaded from doing anything that isn't related to increased productivity or self-improvement? How and why are people subjecting themselves to ever-harder "jobs" for no pay? Indeed, gamers often pay for the privilege of playing, yet their cumulative playtime produces no obvious productive value in the context of how we have traditionally understood our economy. Every hour spent in a massively multiplayer online role-playing game (MMORPG) is one less hour that the player might spend improving their productivity or even pursuing other, more productive play activities, such as running on a treadmill at the gym. So what's going on?

The glib and easy answer, as voiced by many pundits, is that video games are an addiction, a vice just as dangerous to working-class health as the saloons against which moralists agitated over a century ago. In 2006, before he became the British prime minister, Boris Johnson inveighed against games in a

Telegraph column titled "The Writing Is on the Wall—Computer Games Rot the Brain." While staring at the screen with controller in hand, Johnson argued, gamers "become like blinking lizards, motionless, absorbed, only the twitching of their hands showing they are still conscious. These machines teach them nothing. They stimulate no ratiocination, discovery, or feat of memory." Or take *Cyber Hunter,* a game by the Chinese company Netease, which has proven so popular in China—it is not uncommon for people there to average eight hours per day of play time—that the Chinese government has imposed strict time limits on game playing for minors in an effort to curb their purported addiction to these games' "spiritual opium."

But if video games are black holes of ratiocination—if they're vectors for addiction and lethargy—then why do they seem as involved and complex as a detailed hobby or an industrial process? Gaming seems substantially different from other addictive behaviors, such as smoking opium or drinking, for instance. There's nothing very demanding about drinking at a bar. All the task requires is the cash to pay for pint after pint and a backside sufficiently padded to sit on a stool for hours on end. With the exception of quiz night, drinking in public does not require the drinker to solve a series of ever more complicated puzzles. And yet there are world championships for major video games, with multimillion-dollar prizes, which clearly suggests these games are far more challenging and meaningful than rapidly guzzling vodka Red Bulls on a Friday night. (An activity that is definitely a vector for addiction, though not lethargy.)

Participating in a modern gaming environment can be a cognitively complex activity. The best modern games are commitments in a way that cannot be said about other forms of popular entertainment, and they are intellectually engaging in ways that do not conform to what we know about destructively addictive habits. Unlike people who choose to engage in passive leisure during their free time, today's gamers aren't just zoning out: They're locking in.

The amount and variety of cognitive activity required by the virtual environments of today suggest that video games aren't just an escape from the real world: They're an alternative, one that is poised to meet the sorts of psychological needs that are no longer being met by society in its quest for increased productivity. Rather than scorn or fear these virtual worlds that are now commanding so much of our time, we should learn from them, and find ways to incorporate those lessons into the real world.

To understand why video games are becoming ever more complex, and how virtual worlds will integrate with our own, we must transcend this recent model of a society where endless productivity is the ultimate good, and examine the scientific understanding of the raw motivations of human beings. Modern society needs to refocus instead around modes of social and individual fulfillment. Before we can do that, though, let's understand exactly what we mean by *fulfillment*—and why fulfillment leads to productivity, rather than the other way around.

PURPOSE AND FULFILLMENT

In a 2010 interview with *ABC World News Tonight*, the physicist Stephen Hawking asserted that "work gives you meaning and purpose, and life is empty without it." In the context of modern society, there's a lot of truth to that statement. Not only is your job a means of earning the money that you need to live—a meaningful need—but there is an inherent satisfaction in feeling like you are a productive member of society, like you are playing a small part in making society function. As I've established, our modern world is organized around modes of productivity, and we have been conditioned to believe that fulfillment is a byproduct of productivity.

In recent decades, though, the dominant social model of ever-increasing productivity has reduced the scope of meaningful employment opportunities available to workers in the Western world. Many companies have found that they can produce more goods and services for less money if they automate the production process and/or transfer jobs to emerging labor markets. This situation, and the resultant waves of mass unemployment and underemployment, is often primarily characterized as an income crisis. But it is also a crisis of purpose.

Whereas decades ago it was common for people to work at the same company from training until retirement, modern business management practices have decimated the sort of long-term monogamous employment opportunities that were once enjoyed by blue- and white-collar workers alike. As a result, many people today are putting in long hours while juggling multiple jobs and gig work, none of which offer the same sort of steady income and long-term stability as the jobs enjoyed by earlier generations. A 2021 *Guardian* article reported that approximately 4.4 million adults in England and Wales were working for gig-economy companies—a number that had more than doubled since 2016. This trend is mirrored in the United States, too: In 2020 alone, according to *Small Business Trends*, the gig economy in the United States grew by 33 percent, expanding 8.25 times faster than the U.S. economy as a whole. Plenty of other people have no work at all, and spend their days looking for jobs that aren't there and may never return. All of this hustling is hard work. Very little of it is existentially fulfilling.

If work gives you purpose, at least in part, then it stands to reason that when you lose your work, you lose your sense of purpose. As David Graeber noted in his book *Bullshit Jobs,* losing one's purpose can be a psychologically devastating experience. It can make you question your place in this world, and whether there's even any point in going on. In a de-ritualized world that no longer officially values engagement with socially constructed realities, if you can't find purpose at work, it can be hard to find purpose at all.

The ongoing phenomenon of farmer suicides in the United States and elsewhere—precipitated by rising debt, lost income, and, often, forfeiting one's farm and assets as a result of the preceding factors—is but one tragic example of how losing one's purpose in a production-minded society can serve as both a figurative and a literal death knell. "When your farm doesn't succeed or you have to sell off some property, not only are you letting you and your family down, you're letting your family legacy down," a spokesperson for the Ohio Farm Bureau told *USA Today* in 2020. "'My great-grandpa started this farm, and now I'm the one that's causing it to cease?' Boy, that's a tough thought. But a lot of farmers are going through that right now."

This example is not exclusively an income problem. It is also a *purpose* problem, one that is endemic to a society where human labor is increasingly squeezed between ever-rising production demands on one side and ever-improving corporate efficiencies on the other. Productivity tends to increase over the long term, because the machines and programs used in work contexts are improving all the time. But as machine labor becomes more central to production processes, human labor will necessarily become more and more marginalized. And as people lose their centrality of purpose in a society that sees productivity as the primary goal, the current widespread psychological crisis will only get worse.

The solution to this social dilemma lies in disentangling the concepts of purpose, work, and employment—and in realizing that it is just as important to the continued function of society to ensure that people are fulfilled as it is to ensure that they are gainfully employed. Continued employment and ongoing fulfillment are two of the most important things that a society ought to provide to its members. For centuries, Western society has focused on the former, often at the expense of the latter. Digital games and virtual worlds can help rebalance the scales.

While employment can indeed give people a sense of purpose, *purpose* and *employment* are not necessarily synonymous. A paying, productive job isn't the only thing, or even the primary thing, that gives our lives meaning. We know this from our study of ancient metaverses in the first chapter, but we also know this from our own lived experiences. After all, you can work in your garden and derive a great deal of meaning and purpose from that labor, even if it's not an income-producing job and carries no broader social utility. You can work on craft projects, or home repairs, or stamp collections, or any number of other hobbies and pastimes and passion projects, devoting just as much time and labor to them as you would to a paying job, and come away from your efforts feeling happier and more purposeful than you would if you'd spent an equivalent amount of time at your desk.

Likewise, we all inherently know that even the most stable, productive jobs aren't necessarily fulfilling ones. Many jobs are repetitive and boring, replete with recurring tasks that do not really stimulate or challenge the worker. Many workplaces, likewise, do not prioritize employee satisfaction; instead, they take it for granted that productivity creates satisfaction. But I would argue that work is not fulfilling because it's productive. Workers are most productive when their work is fulfilling.

When you feel fulfilled in your work—be it a paying job, a personal project, or something else that demands your long-term cognitive engagement—you tend to work harder and smarter. You seek out new projects and challenges; you are intrinsically motivated to do your job. When you feel optimally challenged, the work can become its own reward.

We know this, in part, from looking at the recent history of gaming. Many of today's immersive digital games are work in the best sense: They are complex cognitive environments that optimally challenge their participants and encourage long-term engagement. People spend hours at a time being productive in gaming environments—getting good at games, accumulating points and skills and accomplishments and in-world status—because these games are built to be fulfilling.

If you stop and think about the times that you've felt burned-out at your job, it's likely that they correspond to times when you haven't felt heard or valued by your managers and colleagues, or times when your tasks have lacked the sort of complexity necessary to make a job feel consistently interesting. In these sorts of disempowering situations, you may well decide to put the minimum effort toward your labors. In his 1975 book *Intrinsic Motivation,* the research psychologist Edward L. Deci described how workers who are micromanaged and deprived of workplace autonomy can rebel against their situations, willfully becoming less productive. "Often, in fact, people will satisfy their intrinsic need to be creative by devising ways to beat the system," wrote Deci. "This may take the form of subtle sabotage and will certainly manifest itself in people trying to get the greatest rewards from the organization while giving the least effort to the organization."

Oddly enough, something similar can happen in games that do an insufficient job of fulfilling their players. It's fair to say, then, that a terrible job is fairly similar to a crappy video game. Good jobs and games lead with fulfillment, then expect productivity to follow. Bad jobs and games lead with production mandates, then expect fulfillment to follow. The crisis of purpose in the Western world today is arguably a byproduct of the Western world spending far too much time stuck in the latter mode, while studiously avoiding any meaningful consideration of the former.

ON MANAGEMENT AND MOTIVATION

For a long time, many big workplaces were organized around specialized job descriptions and strict quantitative measures of productivity, while making relatively few concessions toward employee fulfillment. To these employers, the best way to motivate workers was by rewarding or punishing them for their output, or lack thereof: bonuses and promotions for exceptional performance, demotions or termination for subpar work. The most productive workers received the most external validation, while the least productive workers received the least.

The motivational structure of these workplaces was heavily influenced by an academic field called behavioral psychology, or behaviorism. Among other things, the behaviorists argued that humans are primarily motivated by extrinsic rewards and punishments, and that the best way to get people to do things is to train them to associate certain tasks and behaviors with rewards and punishments. Complete a task, get a cookie. Fail to complete a task, no cookie. Close the deal, get a promotion. Fail to close the deal, get fired. If the boss from *Glengarry Glen Ross* were a branch of psychology, he would be behaviorism.

Behaviorists didn't think that our intrinsic needs were knowable or important, or that humans were even capable of truly autonomous choices. (The arch-behaviorist B. F. Skinner even went so far as to claim that "free will is but an illusion.") Instead, they believed that human beings weren't all that different from Pavlov's dogs, and that human behavior could best be understood as a series of learned responses to extrinsic stimuli. The environment made the man, in other words—and by exerting tight control over the environment, you could make that man into whatever you wanted him to be.

This philosophy, you will have noticed, is not particularly receptive to the premise that constructed worlds of meaning might enrich individuals' lives and create value for society. Behaviorism did not admit the need for or the value of the anti-structural spaces that humans had long chosen to build and inhabit. As the behaviorists' theories gained credence, these other worlds came to seem less necessary to the functioning of the stimulus/response society that industrialization had created.

Radical behaviorism has since fallen out of favor. (Many of Skinner's theories were convincingly dismantled in 1959 by the linguist Noam Chomsky, who argued that behaviorism could not account for the phenomenon of language acquisition in infants, among other things.) But, in its time, Skinner's brand of behaviorism was influential—and, importantly, its tenets aligned with the ever-increasing production demands of the industrial age.

Even before the behaviorists became prominent, many businesses had begun to organize around theories of "scientific management" and efficiency. "Efficient" workplaces became highly organized and controlled environments, in which workers were stripped of their individual volition and trained to perform very specific tasks to very specific standards. These workplaces minimized individual complexity in order to maximize total productivity. As management theories evolved and corporations grew larger and more valuable, many workplaces leaned into behaviorist principles of motivation.

It's interesting to review some of the most popular management books from the twentieth century as a snapshot of what optimal motivation strategies looked like at the time. While some top sellers did indeed encourage managers to consider their workers' intrinsic needs, many more were focused on power and authority, and emphasized structure, measurement, and quantifiable outcomes. Per these books, successful managers were leaders who motivated their employees by telling them what to do and how to do it. These motivational structures were centered around extracting the maximum production from each individual worker. But they also discounted the possibility that workers might have deep-seated intrinsic needs for complexity and fulfillment at work—and that a business might derive significant benefit from situating their workers to meet those needs.

But recent psychological research tells us that our brains are problem-solving organs that need a steady diet of increasingly difficult problems to solve. We want to progress toward mastery in the things that we do. As we approach mastery, we access new and different levels of psychological fulfillment. It feels good to be good at something, and to get better at it over time. Consciously or not, people want to be able to experience continued success—they want obstacles that they can learn to surmount, not ones that will thwart or bore them every single time—which is why it is so fundamentally dispiriting that so many of our jobs seem to be stuck in an endless, enervating loop.

Our days have come to resemble our jobs. Today's world is one in which we are all forever stuck at work, in a society that offers fewer and fewer regular opportunities for leisure or community ritual. Even though our jobs take up more space than ever, they are hardly any more fulfilling than they ever were, because individual fulfillment is not and has never been the point of industrial society. Productivity is the point, and the modern workplace is a tremendously efficient process that has been refined over the years to eliminate all aspects that are extraneous to quantifiable productivity.

Like the behaviorists with whom they share so much, today's workplaces posit that the inner lives of individuals are irrelevant. But as millennia worth of human cognitive engagement with worlds that exist only in our minds have shown, our inner lives *do* matter—and not just in the Jungian or Freudian sense, either, where all of our conscious actions are linearly connected to our dreams and the subconscious. Instead, humans have innate needs for complexity and fulfillment. We want to feel good at things, challenged by things, with a measure of control over our own choices and a sense of belonging within our milieu. We want to use our ingenuity to create value, both for ourselves and for the worlds that matter to us. In *Intrinsic Motivation*, Deci observed that "situations can be structured so that people will motivate themselves. . . . People can be committed to doing their jobs well, and they can derive satisfaction from evidence that they are effective." In other words, the prospect of payment isn't the only thing that might motivate someone to do their work.

This intense desire for intrinsic fulfillment is why so many people today are flocking to digital games and virtual worlds. Today's best games are more than just fun, and they're the furthest thing from being mindless wastes of time. Quite the contrary: They are complexity engines configured to offer the sort of intrinsic fulfillment that so many other social institutions have long neglected to provide. To repeat the refrain from earlier, these games are gymnasiums for the mind, and the more time that you spend inside them, the stronger and healthier your mind becomes.

Of course, there are plenty of games that are bad for you: games that disrespect your time, attention, and intelligence; games that are gratuitously nasty; games that do not even attempt to fulfill any of the players' intrinsic needs. But think of it this way: There are plenty of jobs that are bad for you, too—and these jobs are bad for you in the sorts of ways that people think that *all* games are bad for you. These bad jobs lead to disaffection, alienation, sadness, loneliness, and other negative social outcomes. If you hate a bad game, you can just stop playing it. If you hate your job, well, it's still your job, and often you're simply stuck with it—until you lose it, at which point your crisis of purpose grows even worse.

Games and play reward ingenuity, and as we know from the last chapter, ingenuity within parameters has been central to human psychological health and well-being for millennia. Unlike work environments, many of which are designed to treat employees as replaceable cogs in a production process, games celebrate both outcomes and processes. The best games promote growth, curiosity, problem-solving, teamwork, and creativity. Ideally, we'd want both our games and our jobs to be good for us. But it's easier to make a good video game than it is to make a deeply fulfilling job. Work in an industrial economy will always be oriented around principles of productivity. But games can be tuned to prioritize principles of fulfillment much more easily than a job can.

As a society, we can afford to acknowledge that there is broad merit in positioning people to experience individual fulfillment. We might not be able to completely remake the real world as it currently exists, but we can understand the importance of letting people use their free time to rebalance their lives. As we have done as a species for 10,000 years, we can try to create new worlds founded on more humane principles than our current one. In the next chapter, I'll tell you about the science underpinning today's efforts to create rich, resonant digital worlds that enhance and expand our lives by offering consistent access to fulfilling, useful experiences. Games and play can help people meet their basic psychological needs in ways that their jobs do not. The vast virtual worlds of tomorrow will do it even better.

Chapter 3

BETTER EXPERIENCES FOR BETTER LIVING

I n 1932, the DuPont corporation adopted a motto that captured the zeitgeist: "Better Things for Better Living . . . Through Chemistry." It was a slogan fit for the glory days of the Industrial Revolution. In the post–World War One era, the standardization of manufacturing processes promised to bring people fulfillment through access to more consumer goods than had ever before been available. The things people could buy would be less expensive and higher quality than they had been in previous generations—and new and better things were being invented all the time. In earlier eras, people knew their daily lives would entail certain privations—endless wash-days, cold-weather treks to the outdoor toilet, sweltering homes on summer afternoons—which is perhaps one reason why the afterlife was so prominent in people's minds. But in the modern age, you would no longer have to wait for the next world in order to live your perfect life. Instead, you could buy that perfect life at Macy's.

It was not entirely coincidental that Better Things were presented as a tacit antidote to the ills of the Great Depression. In the 1930s, with the corporate world intoxicated by the gospel of scientific management, some business leaders embraced theories which presumed that motivation, satisfaction, and purpose were all extrinsically derived. The DuPont motto was the product of an industrial ethos that considered inner experience to be irrelevant, one that saw workers as functionally indistinguishable from the machines they operated. As I noted in Chapter 2, some of the most influential organizational and behavioral thinkers of the Industrial Era contended that humans had no inner worlds worth studying or acknowledging. As such, it made perfect sense for DuPont to claim that better things would indeed lead to better living—and, implicitly, that productivity was the point of human existence.

The social structures of the Industrial Era sharply diverged from those that preceded them over the course of thousands of years of human civilization. As I wrote in Chapter 1, human society has almost always had a reverence for worlds that exist beyond our own, and, seemingly, a keen sense of the value that engagement with those worlds can create on Earth. But industrialization rendered those other worlds superfluous to a society that held production and fulfillment to be synonymous. The promise of heaven gradually grew less potent in a world with dishwashers and central air-conditioning.

In recent years, new research in psychology, linguistics, and philosophy has revived the notion of inner and virtual worlds as structures vital to the health of human society—and has reacquainted us with lessons that our ancestors learned long ago. Today, compelling theories of human behavior and motivation emphasize intrinsic factors instead of extrinsic ones. They argue that humans are driven less by the prospect of punishments and rewards than by opportunities to exercise autonomy and ingenuity within a social context. These insights—which, as I will show later in this chapter, are supported by laboratory research—align with what we already know about the ways in which participating in socially constructed realities can promote positive outcomes for individuals and for society. According to this model, people crave a robust variety of engaging, useful, empowering experiences that challenge their minds while connecting them to their peers and the wider world. In other words, they are fulfilled more by the journey than by the ultimate destination. While the prospect of earning a bonus or being fired for nonattendance might keep people coming to work each day, being consistently challenged by jobs that respect their abilities and their humanity is what keeps workers happy.

There are obvious flaws in a social model that presumes that we work to earn, earn to buy, buy to use, and use to discard, over and over and over until we retire or die. An endless supply of new things might serve to juice a developed nation's GDP, but they no longer serve to meaningfully boost its citizens' marginal fulfillment levels. "Things" are the prerogative of our production-centric world. But experiences will be the currency of our virtual societies. As such, I'd like to propose a new slogan to befit our impending virtual age: Better Experiences for Better Living.

There is ample scientific evidence to show us that the most fulfilling lives are the ones that maximize feelings of autonomy, competence, and relatedness. (I'll have more to say on these three factors in a few pages.) These feelings are accessed via activities, challenges, and quests that engage and stimulate the mind. Whether we're traveling the world in order to expand our perspectives, attempting to acquire a new skill or hone an existing one, or simply pursuing a hobby that brings us joy, there is great psychological and practical value in doing things that excite, engage, challenge, and change us. One could even argue that the point of our journey through life is to maximize these sorts of experiences, while minimizing harmful ones.

In a sense, a transporting experience is a form of communion with another world. Human beings seek out experiences—divergences, big and small, from our baseline realities—in order to be changed by them for the better. A big trip to an unfamiliar destination, a visit to an escape room, a wellness retreat: Experiences are opportunities to step outside your daily routine and engage with an unfamiliar set of premises. They offer a measure of anti-structure that can help to counterbalance, contextualize, and change the course of your daily life.

Humans value and pursue experiences not only for reasons of inner fulfillment, but also to learn new skills, prepare for the future, or address some existing issue or problem. The optimal experience, then, is both fulfilling and useful: one that provides intrinsic satisfaction while also catalyzing personal growth and productive change. Constructed worlds of meaning have been providing these sorts of experiences for their participants for thousands of years. By offering people spaces for applied ideation, and new sets of challenges to define, assess, and solve within the context of a social group, these worlds generate fulfillment and value. The purpose of a digital virtual world is to efficiently and reliably create valuable experiences for its participants with greater consistency and fidelity than ever before.

In Chapter 2, I wrote that the best digital video games challenge and fulfill their players by immersing them in complex cognitive environments that offer the sorts of optimal challenges that are scarce in the working world. The digital worlds of the near future will build on the games of today to create a surfeit of rich, useful experiences for participants: quests that are immersive and dense with meaning, in environments that feel like they are teeming with intelligent life. When these digital worlds are put into contact with the real world and one another via digital networks, the ensuing metaverse will serve as a powerful new engine for experience creation, value transfer, and intrinsic fulfillment.

In this chapter, I'm going to talk about useful experiences, intrinsic fulfillment, and how networked virtual worlds will generate the latter by providing the former. I'll show you how psychologists have come to believe that fulfillment and motivation are tied to the pursuit of becoming one's best self, and I'll present some of the research that shows why engagement within constructed worlds of meaning advances this pursuit in ways that our jobs do not. I'll explain why experiences and fulfillment are linked, and why we value experiences not just for how they make us feel, but for what they can teach us. I'll outline some of the ways in which digitally rendered virtual worlds will transform the production of optimal inner experience, as well as the creation and transfer of the value tied to those experiences. And I'll argue that a network of worlds will offer unparalleled access to the sorts of powerful, useful experiences that people used to encounter perhaps once in a lifetime.

Much as the first Industrial Revolution refined the production process for existing goods while simultaneously producing goods that had never before existed, digitally rendered virtual worlds will industrialize the production of optimal inner experiences—but not in the sense of standardizing them and making them cheap. Instead, they will bring newfound reliability and precision to the production of existing experiences while also creating new sorts of experiences that would never have been accessible to us without those virtual worlds. Throughout history, for example, mythology has brought us tales of heroes pursuing noble objectives via fantastic quests. In the very near future, we will all be able to have those sorts of heroic experiences for ourselves—and not just in the superficially interactive manner of modern games. We won't simply "play" at being heroes or villains or generals or pioneer blacksmiths in some fantasy universe. For all intents and purposes, these roles and these settings will actually exist. We will live our own epic tales rather than just reading about other people's grand adventures.

It's up to us to shape this virtual future in a way that's maximally beneficial to individuals and society. If the ideal future comes to pass, it will be because society sees the benefit—the essential humanity—in creating worlds that can consistently engage their residents, rather than treating them as waste products of some hyper-efficient industrial process. We must address society's thoroughgoing crisis of purpose by first understanding how purpose is derived, and by examining why cognitive engagement within virtual spaces has historically served and fulfilled human beings. To do this, let's turn to the fascinating branch of psychology known as self-determination theory.

THE POWER OF SELF-DETERMINATION

If you could do anything in the world right now, at this very moment, what would you choose to do? Maybe you love video games, and your ideal day consists of exploring all facets of a richly rendered digital world. Perhaps you're at your happiest when walking the beach with a metal detector, scanning for treasures buried under the sand. Maybe you really like to read, and you'd prefer to be sitting down with an engrossing new book on the coming age of virtual society. (If you relate to that last example, then congratulations: You're already living your best life.)

We can all probably think of some activity that we find consistently challenging, stimulating, and exciting—some task or pastime that makes us feel good at life and happy to be alive. Our physical survival doesn't generally depend on these sorts of activities, we don't primarily see them as pathways to fame and fortune, and no one will yell at us if we leave them undone. We simply know that we need to do these things in order to be truly happy. As the psychologists Edward L. Deci and Richard M. Ryan have argued, pursuing and fulfilling these sorts of intrinsic needs is the key to sustainable motivation and psychological well-being.

Deci and Ryan are the foremost proponents of a branch of psychology called *self-determination theory,* which claims that our intrinsic needs are primal needs, ones that are critically important to our ongoing health and happiness as a species. Intrinsic fulfillment is not usually contingent on any external rewards, such as payment or praise; instead, the joy derived from pursuing these needs has less to do with what you will get after you meet the need than with what you get out of the process of meeting it.

Over the past forty years, in a series of influential books and papers, Deci and Ryan have argued that situating people to pursue intrinsic fulfillment is the best way to help them grow as individuals and take control of their own lives. "The term self-determination refers to a person's own ability to manage themselves, to make confident choices, and to think on their own," Deci and Ryan wrote in their seminal 1985 text *Intrinsic Motivation and Self-Determination in Human Behavior.* A self-determined person, they said, was more likely to also be a happy, healthy, and well-adjusted person. The more well-adjusted people there are in a society, it stands to reason, the more stable that society becomes.

Since *Intrinsic Motivation and Self-Determination in Human Behavior* was first published, self-determination theory has come to offer a compelling framework for how and why we might center our lives and our societies around principles of fulfillment. But what, exactly, does fulfillment mean in this context? And how does it relate to the individual and community uplift we feel when we engage in virtual worlds?

Deci and Ryan argue that humans are motivated by their fundamental needs for *autonomy, competence,* and *relatedness:* the building blocks of human fulfillment and psychological growth. *Autonomy* is, basically, the desire to set your own agenda: the freedom to articulate and pursue your own goals and projects while exercising control over your own behavior. *Competence* is the need to feel good at things: to be able to grow in knowledge and accomplishment, to acquire and master skills and deploy them in a range of environments. *Relatedness* is the need for a sense of connection with the people around you: those feelings of attachment and belonging that can make you feel included in and uplifted by group dynamics.

Fulfilling these three needs correlates with positive mental health outcomes. When we feel autonomous, competent, and related, we tend to feel happier and more motivated. We interact better with the world and the people around us. We sleep more soundly at night. And we're more likely to pursue a path of self-determination that can help us become our best selves.

Self-determination theory maintains that humans are organisms that seek out complexity, and that our brains are problem-solving organs that require a steady diet of increasingly difficult problems to solve. When we successfully solve hard problems, we feel capable and happy. When we are deprived of hard problems to solve and master, then we tend to stagnate. As I noted in Chapter 2, today's best digital games are engineered to provide complex challenges that meet people's inner needs, while also being fun to play. Within a gaming context, the dual imperatives of fun and fulfillment are linked—which is one reason why some people have trouble taking games seriously.

In a world increasingly focused on productive uses of time, we're conditioned to be suspicious of things that are just "fun," without holding any apparent broader utility. But as games have evolved, so has game developers' understanding of exactly why people find their games to be so enjoyable. In their 2007 paper "The Player Experience of Need Satisfaction" and in several subsequent publications, Scott Rigby and Richard Ryan have suggested that, in gaming environments, fun is best understood as a byproduct of need satisfaction.

The best games satisfy players' needs for competence by offering them both an ongoing series of *optimal challenges*—that is, challenges that are matched to the player's skill level, and that get harder as the player improves—and *"mastery in action"* experiences, in which talented players can show off their skills and navigate in-game challenges without breaking much of a sweat. Games can satisfy players' needs for autonomy, according to Rigby and Ryan, by offering them a range of *opportunities for action* and the ability to construct their own in-game identities. Finally, games can satisfy players' needs for relatedness by offering opportunities to interact, collaborate, work with, and learn from other characters within the game.

In their book *Glued to Games,* Rigby and Ryan suggest that the text of a game is less relevant than its psychological subtext in understanding why it does or does not succeed—much as the specific content of a community ritual is often less important than the social function it fills. There is nothing inherently fun about facing off against a zombie horde, for instance, as players have done in dozens of popular games over the past two decades. If we were to encounter such a horde in real life, we would likely run away as quickly and as quietly as possible. In a multiplayer gaming environment, though, killing zombies feels fun because doing so satisfies the player's core needs for autonomy (you choose which zombies to kill and how to kill them), competence (you get better at killing zombies over time), and relatedness (you often kill zombies alongside the other people in your zombie-killing guild).

Critics might look at a violent game and see a "murder simulator" with no redeeming value, a game that is teaching children to kill. But the joy in the game isn't in the text of the game—the guns and the hordes and the splattering blood—but in the subtext, in the way the game consistently presents its players with situations that make them feel capable and satisfied. Though critics might say that these games teach children to associate violence with satisfaction, I think it's more accurate to say that they teach people to associate problem-solving with satisfaction.

It's important to note a fascinating paradox here. On the one hand, I know of no hard evidence accepted by the scientific community that has ever conclusively linked simulated violence to actual violence. People seem completely able to distinguish between fantasy and reality in that regard. Yet at the same time, the intrinsic impact of these virtual experiences is very real. The tangible fulfillment offered by virtual experiences appears to be just as psychologically nutritious as any "real" experience—in some cases more nutritious, due to the unrestricted nature of the challenges and opportunities available in a virtual world. Winning in a video game has the same impact as winning in any other activity, in terms of the feelings of fulfillment and competence derived from the victory. Relationships forged inside games, too, can be real relationships, with the same sorts of tangible benefits as real-world friendships. Make no mistake: Virtual experiences *are* real experiences.

As studios and programmers work to build ever more complex and immersive games and virtual worlds, self-determination theory has come to greatly influence the games industry. Some studios retain self-determination theory experts as consultants, where they work with developers to help them build environments that offer boundless opportunities for users to experience feelings of competence, autonomy, and relatedness. These fulfillment-rich digital environments point our way toward the future while also hearkening back to the lessons of the past.

As I wrote in Chapter 1, many preindustrial societies instinctually knew the social value of tending to people's inner lives. They did so via socially constructed realities that became important through participation and belief, often accessed by ritual experiences that inverted the structure of daily life while creating *communitas*. Our post-mythological era has moved away from these worlds, but the human hunger for optimized inner experience still remains. One of the takeaways from self-determination theory is that fulfillment is not optional when it comes to mental health. Its absence creates serious problems for individuals, and its presence serves a function similar to exercise: a near-universal good with benefits that extend to every aspect of life. It is thus unsurprising that many people who find themselves consistently unable to find fulfillment in the real world are gravitating toward games in their free time.

If games can provide their players with intrinsic fulfillment, then expansive, networked virtual worlds will be able to boost those fulfillment levels and convert them into social value. These worlds will be better equipped to provide their participants with meaningful experiences. A virtual world differs from a game in a few ways, key among them being that a virtual world is a consequential place. The things that happen in that world matter to the participants, much like they would matter if those things happened in real life. If your virtual house gets burgled in a meaningful virtual world, for example, you may well find it as devastating as you would if your real-world house were burgled; in any case, the things that were taken from you in virtual space won't magically reappear if you log out and then log back on.

In a meaningful virtual world, actions have consequences that persist betwixt and between sessions. This continuity of consequence is a good thing. People *want* to live in a world where their actions matter. Imagine a virtual world in which your avatar picks up a rock and heaves it through the window of your neighbor's house. If this world is a meaningful one, then there'll likely be hell to pay. Your neighbor would come out and yell at you. A fight might break out. The cops might come. You'd have to pay for the damage. You'd develop a negative reputation on your street, and your relationship with your neighbor might not be the same for a long, long time. Things wouldn't go back to normal the next day. In this example, it doesn't really matter if the rock you throw resembles a photorealistic geode. The fact that your virtual neighbor comes out of the house, reams you out for breaking the window, and then hates you for years thereafter is what makes the whole situation feel real.

Virtual worlds' ability to manage and manifest complexity while measuring the satisfaction and growth that participants are experiencing therein will lead to new frontiers in individual need fulfillment. Achieving this type of data-driven fulfillment is already very possible. The games industry has established, effective ways to measure engagement, and it's easy to imagine these methods evolving and improving in more complex universes. Another differentiating characteristic between even the most advanced games and the virtual worlds of tomorrow will be size. Assuming feasible improvements in hardware and software, a capacious virtual world will conceivably be able to serve millions of simultaneous participants—which would be orders of magnitude more populous than even the most sophisticated current multiplayer games. This population density will create network effects that will be felt most keenly in the quantity and quality of experiences available to people within the world.

BETTER LIVING THROUGH BETTER EXPERIENCES

The monomyth, as articulated by the writer and comparative mythologist Joseph Campbell, is the archetypal hero's journey. In it, a hero goes on an adventure, learns things, faces challenges along the way, and, in the end, is meaningfully changed by the experience. From Odysseus traveling to Ithaca to the Buddha's quest toward enlightenment to Luke Skywalker's arc in the *Star Wars* films, the hero's journey has been a building block of contemporary culture and belief systems for millennia. These sorts of stories allow us to live vicariously through their protagonists, and to imagine what we might do if we found ourselves in their place.

Though many heroic tales are rooted in oral tradition, once they are written down or filmed they tend to become fixed entities. George Lucas's periodic tweaks to the source material notwithstanding, the *Star Wars* films do not materially alter from one viewing to the next. You can watch *Star Wars: A New Hope* a hundred separate times, memorize all the dialogue, and feel emotionally connected to the characters, but the film will not evolve to match your increasing mastery of it. It stays the same even as you grow and change, offering you the same experience over and over again. Indeed, the hunger for new experiences within the *Star Wars* universe is what animates the creation of so much *Star Wars* fan fiction.

Likewise, passive engagement has long been the predominant form of engagement with monomythic media. Your role in the tale was limited to that of a spectator. Although oral storytellers inhabited *The Odyssey* each time they retold it, listeners— and, latterly, readers—could not physically put themselves in Odysseus's place and experience the thrill of escaping the Cyclops, navigating Scylla and Charybdis, or matching wits with Circe. Reading about someone else's heroic journey can be thrilling in its way, but, from the standpoint of the potential psychological fulfillment one might derive from the experience, it just isn't the same as having your own adventure. This limitation applies to many other kinds of stories, fantasies, and experiences in our culture. Most of our cultural touchpoints are not particularly interactive. No matter how much you may want to, you cannot experience what it's like to go to Hogwarts: a painful lesson that every eleven-year-old—including, once upon a time, this author—must eventually learn.

For decades now, digital games have been sending ordinary people off on the sorts of potentially transformative adventures that Campbell would consider canon. "A hero goes on a quest" is effectively the log line for most digital games today. In the process of advancing through a game, players learn new skills, encounter people and information that aid their journey, and come to understand the world of the game better than they did when they first began. Within the context of MMORPGs, which might feature thousands of other humans inhabiting the same adventure space, games can start to feel like entire worlds. Digital games are optimized to provide their participants with the sorts of positive experiences that let them become, in a sense, the heroes of their own stories. Genres such as real-time strategy games, simulation games, and even flight simulators provide a wealth of other sorts of potential experiences. Admittedly, these games are generally limited in their interactivity; even the most advanced games of today are ultimately simple universes when compared to the real world.

Even with their limitations, today's games are an improvement on the sorts of heroic adventures available in the real world, if only because it is actually very difficult to have heroic adventures in the real world. But just as the rise of digital games expanded access to entertaining experiences, virtual worlds will offer access to dramatically more consequential experiences. If the advent of digital games offered the world a new paradigm for experiencing heroic journeys, then virtual worlds will make those journeys matter. Those journeys will be situated in living, breathing worlds of real importance to society. These worlds will offer their users unprecedented opportunities to experience feelings of competence, autonomy, and relatedness.

For an experience to matter, you must be able to have control over a broad range of decisions and options, rather than have your decisions artificially limited by the parameters somebody else has imposed on your quest. You must also feel the sort of relatedness that comes from true investment in the stakes of the world in which you exist. You may even be co-creating these worlds through your actions, in ways that significantly build on the more primitive opportunities for creation available today in games such as *Minecraft* and *Roblox*. You may even earn your income doing so—a prospect that I will discuss in Chapter 7.

Besides the fulfillment of intrinsic motivations, there is a second dimension to what can make an experience valuable: the skills, lessons, and perspectives we take away from it. When students go off to college, for example, they certainly hope that they will enjoy their time there. But most of them also hope that they will grow in knowledge while learning about themselves and the world, and that those lessons will serve them well for the rest of their lives. Experiences can both yield commodities and function as commodities themselves. They are treasures of the mind that can not only fulfill us but change us into something better, including increasing our earning potential or our social status.

There is often nothing immediately productive about a useful experience, at least by the dominant standards of the Industrial Age. To the contrary: Useful experiences generally consume time and attention that you might otherwise devote to your work. If you go off to college to study agricultural management techniques, that choice generally means that you won't be home at the family farm when they need your help with the harvest. But if experiences sap your productivity in the short term, they tend to make you more motivated and productive in the long term.

Travel is a classic example of an experience that is not directly productive yet is understood to be a worthwhile use of time. The concept of the life-changing youthful sightseeing adventure is rooted in an Enlightenment tradition known as the Grand Tour, in which wealthy young people would mark their entry into adulthood by spending months or years immersing themselves in all the art, culture, and history that Europe had to offer. The rationale was that there was great intrinsic value in seeing and experiencing the wider world, and that the long-term value of taking time off to travel exceeded the short-term value of spending that time in more directly productive activity.

While the whole premise of the Grand Tour is a pleasant one, the fact remains that there was only ever a vanishingly small number of humans on Earth at the time who could afford to spend months or years away from home exploring Europe in hopes that they might be somehow improved by proximity to great art and Gothic cathedrals. Technology has historically served as an experiential equalizer. Over the past 150 years, improvements in travel technologies have made the Grand Tour accessible to anyone who can afford a discount airline fare and a hostel bunk. Modern travelers can now see the same sights as their wealthy predecessors for a fraction of the time and cost that such a journey used to require.

Just as the rise of cheap flights and ubiquitous hostels improved and democratized the European travel experience, digital technologies have long been used to speed and streamline many other sorts of time- and labor-intensive processes that serve to separate people from potentially useful experiences. Whether we're trying to learn something new, download a bunch of books and papers, or simply seek out opportunities to communicate and connect with old friends, we're all accustomed to using computers to achieve a broader spectrum of high-quality, reliable experiences than we might otherwise be able to access in the outside world. The same will be true within the virtual worlds of tomorrow.

The purpose of virtual worlds is to efficiently and reliably create fulfilling and useful experiences for their participants. Their unique ability to process and manage complexity will allow them to produce these sorts of high-quality inner journeys with precision and regularity. The key to maintaining quality standards will lie in our ability to measure and quantify the relative fulfillment and value of an experience in one virtual world, compared to a similar experience in another. The relevant metric here isn't just attention, but engagement, over both the short and the long term.

Eventually, as virtual worlds become not just complex and lively simulations, but also environments that we can access through ever-richer interfaces, they may begin to not only match but actually exceed the fulfillment possible in the real world. In order to understand the progression of experiences within this projected evolution, I'd like to touch on the concept of *near and far experience.*

NEAR AND FAR EXPERIENCE

The experiences that I call *near experiences* are those that are available to us right now, or will be available to us soon, based on the technologies that we currently have or are immediately on the horizon. What I call *far experiences* are those that we can credibly expect to have at some point, based on the expected evolution of the technologies we use to create and access those experiences. We can compare the "near internet" of twenty-five years ago—a largely homemade pastiche of personal websites, low-resolution graphics, good intentions—with the sleek, slick, hyper-functional internet of today as a way to understand this concept. We're now at the far experience of the internet, and while, from a social-impact perspective, the internet may well have devolved over time, the quantity and quality of the possible experiences now available there have expanded so much that today's internet barely feels related to the one we started with.

The near experiences in today's virtual worlds and massively multiplayer online games are very different from the ones that will eventually be made available to us in the far future. But just as the early internet often felt magical and meaningful to its users, the near experiences of today's virtual worlds are likewise transformative in their way. Even now, people can have social and educational experiences in virtual worlds that improve on the equivalent opportunities available in the real world—while also improving on the opportunities available in the far internet to which we're all accustomed today.

Take, for example, the process of making new friends and building new relationships online. There are plenty of stories of people who have never met in person becoming real friends based exclusively on the exchanges they've had over time on the internet. Shared experiences create a baseline common ground upon which it becomes easier to build a meaningful relationship—and virtual worlds can facilitate this process much more effectively than social media or video games. Even interactive social environments in games are usually limited to just a handful of players, who often cannot simultaneously speak or interact with the world in complex ways. (These sorts of limitations are largely technical ones, and I'll discuss them in greater depth in the next chapter. It's worth noting that newer technologies, such as the M^2 metaverse platform, can allow tens of thousands of players to congregate together.)

Research shows that sharing a memorable experience with a stranger is a reliable way to create a lasting bond with that person and turn a stranger into a friend. In the book *Friendship Processes,* Beverley Fehr taxonomized four "dyadic variables" in friendship formation, the first of which was "companionship (e.g., sharing an activity or experience)." In a 2004 paper in the journal *Personal Relationships,* Barbara Fraley and Arthur Aron found that having a humorous shared experience led to an increase in the chance of feelings of closeness to one's interaction partner. Writing about friendships within virtual worlds in a 2011 paper in the journal *Ethics and Information Technology,* Nicholas John Munn argued that "to the extent that shared activity is a core element in the formation of friendships, friendships can form in immersive virtual worlds as they do in the physical world."

Virtual worlds will industrialize the production of shareable experiences, and thus enhance the friend-making process. Online games can have siloed communities focused on playing a single game. Though friendships can and do arise there, they are often limited to their in-game modes of expression. Thriving, connected virtual worlds that are part of a metaverse will create dramatically more contexts in which people may meet, have credible reasons to form long-term relationships, and pursue and evolve those relationships. You may meet a celebrity in a stadium of thousands in a virtual world and have the chance for a fascinating one-on-one interaction involving teleporting to some other world for an adventure that would be impractical, implausible, or unsafe in the real world. It might even lead to a virtual job working with your idol. Many kinds of real-world social interactions are potentially better suited for virtual space, if only for reasons of convenience. A political rally for a major international movement could involve fans and supporters from all over the world congregating instantly. A global university campus might exist, mixing students from everywhere in a context free from many kinds of discrimination.

The near experience of learning in virtual worlds will also improve upon equivalent experiences on the internet and in the real world. In *The Matrix*, Keanu Reeves's character, Neo, having just escaped the titular computer simulation, downloads a bunch of fighting techniques directly into his brain. He does so by means of a direct neural link with a computer, which imparts to him instant expertise in jujitsu, tae kwon do, kung fu, and countless other combat methods.

In a world without advanced brain-computer interfaces, it's unlikely that learning kung fu—or any other skill—will ever become as easy as pushing a button. But the basic premise of the scene isn't all that far-fetched. At the very least, it's a metaphorical representation of the ways in which a computer-aided learning experience can differ in both scope and velocity from its real-world counterpart. In the real world, it might take decades to study, meditate, and train long enough to truly master kung fu. In the world of *The Matrix*, though, Neo was able to simulate a decade's worth of kung fu training and condense the learn-by-doing process into an incredibly short period of time. The computer didn't just simply download a bunch of kung fu textbooks into Neo's brain: It ran him through innumerable simulated kung fu matches in rapid succession until he had accumulated enough experience to be said to have mastered the art.

There are real-world parallels to this sort of rapid skill-formation. Take the notoriously challenging game of Go, for example. The British company DeepMind developed a product called AlphaGo that was able to master the game so thoroughly that it invented new strategies that stunned the world's best players. AlphaGo developed this knowledge in part by playing against itself in vast numbers of games in accelerated time, learning from these experiences. This process is a vastly more efficient version of the ways in which human players learn from their mistakes through useful experiences. Access to a meta-verse of simulated worlds will open up thousands of possible learning environments in which we might dramatically improve real-world performance in many skills.

As these examples show, strategies and skills that would take a long time to develop in the real world can emerge very rapidly from simulations. While simulations can help people learn faster, they also can allow individuals and groups to engage in wholly novel situations and experiences, and to learn from them, so that they might be better prepared when similar situations arise in the real world. Models and memories of past experiences are the basis of learning, which means that an engine that generates experiences is the ultimate learning tool.

There are countless practical applications for learning mediated by a powerful digital experience engine. Let's take war and military tactics as an example. For millennia, military training and strategic planning have been imprecise processes. You can put a soldier through boot camp, and you can put a platoon through a war-gaming exercise, but the fact remains that a training environment, both physically and psychologically, will almost always be different from the environments found in actual combat situations. Likewise, though military strategists can formulate battle and contingency plans, it's also basically all just educated guesswork. As the military aphorism goes, "No plan survives first contact with the enemy." There's no way to really know whether a plan will work, or exactly how or if the plan is flawed, until it's deployed. Then, if and when the plan doesn't work, it's usually too late to scrap it and devise a better one.

In the real world, militaries have been reliably slow to evolve their tactics. Military history shows us that when an army encounters a new situation for which it isn't prepared—new and unfamiliar terrain, newly powerful weapons, an opponent with unorthodox combat tactics—it loses horribly. During the First Samnite War, for instance, the Romans had used the Greek phalanx system to fight their battles, only to find it unsuited for the rocky hills of Samnium. It took a series of defeats during the Second Samnite War for Rome to switch to an entirely new system of field organization, called the maniple. This system was far more maneuverable, Rome was far more successful, Samnium was eventually brought under Roman control, and Rome went on to conquer most of the known world. But the lessons of the last war aren't always still relevant by the time the next war comes around. Instead, there's often a new opponent, with new and unfamiliar tactics, and the cycle of meeting the next war with strategies from the last one begins anew.

This cycle happens in part because militaries just don't fight all that many wars, which means that they just don't accumulate the requisite experience to get better at adjusting their strategies for optimal results. As described in *Empire of the Deep*, Ben Wilson's brilliant book on British naval history, European navies took hundreds of years to progress through different tactical iterations just to realize that simple strategies, such as lining up "ships of the line" and firing in unison, would outsmart intricate plans made impossible to execute by the confusion of naval warfare. Military strategies are a matter of trial and error, and if you don't have all that many trials, or if unsuccessful trials result in your death, then it will take a very long time to minimize or eliminate errors.

Virtual worlds will make it a lot easier for militaries to speed up their learning processes and prepare for combat situations that haven't happened yet, rather than relying on guesswork, intuition, and old lessons that might no longer be relevant. Today, thanks to pioneering work in the metaverse, led by a few companies (including, I should selfishly add, my own), militaries can use virtual worlds to simulate combat scenarios over and over again, much like AlphaGo simulating millions of games of Go before facing off against the world's best players. They can pit soldiers against insurgents in virtual terrain, play those battles out again and again, and gradually extrapolate lessons from the results; those lessons can then be applied to real-world situations, as a means of flattening the learning curve and attaining better results on the ground. The military aphorism about no plan surviving first contact with the enemy becomes less relevant if you can test your plan a million times in virtual space before deploying it in real life. These technologies now extend to massive-scale simulations of entire nations. It may be that, in the future, the most powerful militaries will be those with the ability to see around metaphorical corners through access to vast hypothetical realities. In William Gibson's *The Peripheral*, he envisions a scenario in which new weapons are discovered through observing a pocket reality engaging in World War Three.

Both near and far experiences in virtual worlds will have the opportunity to be as purposeful and meaningful as an equivalent experience in the real world. In time, once the metaverse reaches a critical mass, we will encounter more and more situations where a metaversal experience of something is quantifiably and reliably more fulfilling and useful than the best possible version of its real-world equivalent—where people take more value from virtual-world experiences than they do from real-world experiences. While it might be hard to imagine humanity ever crossing this threshold, I'll refer you back to the ancient metaverses I described in Chapter 1 to note that we have vast experience with characterizing other worlds as better or more meaningful and fulfilling than our present world. Once we create a fulfilling other world that people can actually visit, then that world may well start to eclipse our current world in terms of the value that people assign to it.

As we've seen over the past hundred years, the industrial economy is not set up to produce individual fulfillment. The presumption that happiness is tethered to modes of production and consumption is both flawed and unsustainable. A virtual world that is focused on individual need fulfillment will eventually be seen as preferable to one that is increasingly unable to acknowledge, let alone meet, its residents' needs. The world that delivers the best experiences will be the one that holds the most value.

By *value*, I'm referring not just to intrinsic fulfillment, but also to social and tradeable economic value. Throughout human history, great value has accrued to the physical artifacts and social relationships produced by meaningful experiences. The same will be true in the metaverse. With a surfeit of meaningful metaversal experiences will come an equivalent level of goods and services: memorabilia, artifacts, reputation, relationships, and so on. The recent NFT craze indicates the value that can accrue to purely digital artifacts. In the age of virtual society, NFTs will be totemic: symbolic digital representations of meaningful experiences within a virtual world. NFTs in virtual space will allow us to securitize our most cherished memories, to count them as the assets that they are and have always been. The accumulated value of the goods, services, experiences, and artifacts within a virtual world will create economic and cultural power that will matter both inside and outside the metaverse, and will bring with it all of the good and bad that those forces entail.

At the beginning of this chapter, I spoke of the industrialization of inner experience in the context of the metaverse. As with the real world, industrialization comes with the real risk that the accompanying profits and power will consolidate into a few hands, and that those hands will be the wrong ones. While I'll have more to say on this topic in the second half of the book, I will note that in an optimally valuable metaverse, if one world stops providing optimal fulfillment, then you'll be able to adventure on to a new world that more directly meets your needs. Competition between realities might even serve to drive better opportunities for all. If you're feeling stagnant or stymied at home, then you can head off on a sort of Grand Tour of your own, in order to benefit from all of the new perspectives and opportunities that that sort of experience can provide.

The currency of virtual worlds will be in the quality and quantity of the experiences they provide. The worlds that do a good job of meeting their users' experiential needs will thrive, and the ones that do a bad job will ultimately fail. In his book *Italian Journey,* the German writer Johann Wolfgang von Goethe explained that the purpose of his own extended European trip was to "discover myself in the objects I see." Seeing marvelous things is just where the metaverse will begin. Networked virtual worlds will be filled with objects that you can see, feel, touch, hold, and use. There will be more to see, do, experience, and think about in the metaverse than we can begin to imagine today. And while advanced graphics will play a role in delivering these experiences, the paradigm shift I'm talking about has very little to do with graphical complexity. What matters here isn't the prospect of better, more complex graphics—it's the prospect of better, more complex interactions. In the next chapter, I'll explain why.

Chapter 4

A FRAMEWORK FOR COMPLEXITY IN VIRTUAL WORLDS

The 2018 film *Ready Player One,* based on the 2011 novel of the same name, centers around a virtual world that exists on the far edge of far experience. The virtual world is called the OASIS, and it is basically all things to all people: a shopping mall, a library, a social commons, a job site, and a locus for adventures of all sorts. Advanced graphics and haptic technology make the OASIS a fully immersive environment. In it, countless simultaneous participants can pursue pretty much any activity, from the fantastical to the prosaic, with no lag, gaps, or glitches. The OASIS exists in counterpoint to the "real world" of *Ready Player One,* which is depicted as a wasteland from which people are desperate to escape.

Ready Player One is far from the only work of fiction to posit a future in which complex digitally rendered virtual worlds co-exist alongside the real world. From films such as *The Matrix* and *Johnny Mnemonic* to books such as *Neuromancer* and *Snow Crash*, popular culture has consistently depicted virtual reality as a medium that will be indistinguishable from or even superior to actual reality. But the fictional disparities often depicted between these rich, fulfilling virtual worlds and the barren, desolate outside world also lead people to interpret these digital simulations as either control mechanisms or escape fantasies—modern-day "bread and circuses" created to distract participants from the dreariness of their daily lives.

These fictional worlds are rarely meant to represent any sort of idealized future for humanity. Rather, they are representations of our flaws as a species; plot devices in morality tales that present futures in which our vanities lead us to instantiate hell while calling it heaven. The real world inevitably degrades in these virtual futures because the virtual worlds of fiction are presented as destructive ones built for dubious purpose. No value transfers out of them because, ultimately, nothing valuable happens within. Together, these worlds serve as a warning and a prediction that creating and inhabiting complex virtual worlds will only serve to harm humanity. When this future comes to be, authors and filmmakers have cautioned, it will be a sign that our species has passed the point of no return.

For most of us, our current experiences of and familiarity with virtual worlds are still mediated by their fictional representations. Books, films, and television don't just give us mental models for virtual worlds: They also influence the language we use to describe them. Neal Stephenson's book *Snow Crash*, for example, coined the term *metaverse* and popularized the term *avatar*. In his story "Burning Chrome," William Gibson both neologized the word *cyberspace* and defined it as a network-aided "mass consensual hallucination." Where's the lie?

In many ways, these fictional works will have helped to bring the digital metaverse into existence, by sparking the imaginations of the thinkers and developers who are helping to build the virtual worlds of today and tomorrow. But these works are also responsible for a lot of the misconceptions about the metaverse: what it's for, what it will do, and what it will mean for the outside world. While works of cyber-fiction have helped us to visualize the metaverse, they have also served to preemptively taint the future by imbuing it with connotations of sin and decay.

The actual digital metaverse won't be a plot device in a science-fiction film. And, as I wrote in Chapter 1, these worlds have never stood wholly apart from our own world. To the contrary: Virtual worlds have historically created value for the real world. The impending digital versions of these worlds will follow the same imperatives as their predecessors. The social goal of these worlds is to create useful, complex experiences that make the real world better, not destructive ones that make it worse. In order to get from here to there, though, we must examine the models we've been working with, so that we can separate fact from fiction and light the way toward a usefully complex future. Complexity will be this chapter's key theme.

We are still a ways away from developing virtual worlds that can compare with even their most rudimentary fictional counterparts. While it is easy, thanks to the movies, to imagine a seamless metaverse of infinite complexity, it's quite a bit more complicated to build one. In the gaming industry, games that could credibly be called virtual worlds have only just begun to arrive on the marketplace, and they usually leave much to be desired. Today's virtual worlds can generally host only a tiny handful of players engaging in very basic activities.

If you aren't a gamer, you may find this statement confusing, especially given that, at least superficially, games have become much more beautiful than ever before. A "game engine revolution" has meant that companies such as Unreal and Unity have built wonderful front-ends for game experiences, with easy-to-use tools to produce great graphics, animation, and user interfaces. These front-ends have been combined with the rise of dedicated graphics hardware that makes it computationally feasible to render hyper-realistic environments in real time. However, these advances have not been matched by equivalent progress in networking, simulation, and back-end or "server-side" improvements that would allow virtual worlds to bring actual life to these pretty graphics.

At maximum capacity, products such as Unreal Engine can support not much more than a few hundred people in a rich virtual world before the limits of running an entire simulation on a single computer cause things to fall apart. When you hear about universes such as *World of Warcraft* having millions of players, what this means is that the players are distributed across many identical copies of the world, and thus are unable to all interact at the same time. This "sharding" dramatically limits the consequentiality of interactions with and within these worlds. There is little point in being the greatest hero in a world with just a handful of people in it. As later chapters will show, the metaverse absolutely requires tremendous scale in order to achieve its value proposition.

Ironically, while they correctly imply the enormous technical capabilities that would solve all these problems, and even describe possible solutions in surprising detail (such as the distributed simulation engine alluded to in *Ready Player One*), fictional representations of digital worlds tend to misapprehend the point and purpose of the metaverse, which is to facilitate on-going value transfer from one world to another. Instead, inhabitants of these fictional virtual worlds tend to fixate on either escaping them, controlling them, or both. The plot imperatives of fictional works can also lead creators to depict their virtual worlds as games that players can win, which is itself a bit of a confused concept. Only in fiction will the metaverse be a zero-sum game.

The pessimistic vision of virtual worlds as addictive environments people use to escape their problems bears little resemblance to the metaverse that I see us building in real life. To the contrary, by providing its users with psychological fulfillment and useful experiences, virtual worlds will improve the outside world by giving users the tools they need to thrive in their daily lives. Virtual worlds will create value for the real world, and a metaverse is a conduit for that value.

The gap between the virtual worlds we have now and the ones that we envision in fiction serves to perpetuate skepticism and misunderstanding. The fictional metaverse creates unrealistic expectations for the real one, and can generate inaccurate assumptions about what needs to happen for today's virtual worlds to feel "real." Likewise, fictional models can lead critics and moralists to preemptively dismiss the prospect of virtual society as frivolous, worthless, or even dangerous. But the only valueless virtual worlds will be those that are inspired by surface impressions of their fictional models—worlds that emphasize visual immersion and hyper-realistic graphics while minimizing useful experiences and opportunities for intrinsic fulfillment. These worlds will not be usefully complex ones. We must move beyond these models if we are to create an optimal future.

In real life, as in fiction, many people still conflate graphical immersion and useful complexity when evaluating the quality and utility of virtual worlds. According to many of these parties, realism in the context of digital environments is broadly synonymous with *photorealism*. Do the world's avatars look like actual people, not cartoons? Do its backgrounds resemble the sort of visuals you might see in a live-action film or television show? When the wind blows in the context of a virtual world, can you see an avatar's hair ruffle?

The worlds that can provide affirmative answers to these questions and other, similar questions are often the ones that are thought to be the most complex—and, in terms of the computational power used for graphics, they are. But this is a very narrow definition of complexity, and it isn't particularly helpful when trying to understand the value that the metaverse will create. The complexity manifested by these visually advanced worlds does not always lend itself to the kinds of valuable experiences I've described in previous chapters. As the history of virtual reality and gaming demonstrates, the utility and value to the user of a virtual environment come less from the fidelity of the visuals than from the variety, complexity, and usefulness of experiences the environment contains. We should judge the utility of these worlds based on the complexity of their environments and how well they can fulfill fundamental motivations.

In Chapter 3, I explained that virtual worlds are vehicles for self-determination and psychological fulfillment. To begin to understand the *how* and *why* of the metaverse—the optimally valuable metaverse that I see, not the empty ones of corporate blandishment and dystopian fiction—you must first understand the concept of complexity in the context of virtual environments.

In the rest of this chapter, I'm going to explore complexity as it relates to the worlds linked within a digital metaverse, in order to help separate virtual reality from rhetoric. My goal is to give you the tools you'll need to differentiate between the levels of complexity we can manage today and the levels of complexity required to make some of the bolder predictions about virtual worlds come to life. I'll begin by taking a brief look back at the first wave of embodied virtual worlds, in order to explain why great interactions matter more than great graphics when it comes to building a sustainably complex virtual environment. I'll introduce the concept of *useful complexity* in order to show you why depth is so important if your goal is to build virtual worlds that can fulfill individual and social needs. I'll explain the technological challenges involved in ascending from one level of complexity to the next, and why "communications operations per second" is the most important heuristic to bear in mind when evaluating the complexity of a given world, or any proposed infrastructure that runs such a world.

Eventually, all of this complexity may mean that virtual worlds will end up feeling more real than the real world. We're not quite there yet—and not just because today's technology cannot yet render humanoid figures lifelike down to the last detail. Though many game developers today work with experts in self-determination theory, many others continue to prioritize user immersion over user fulfillment; they have chosen to design from the outside in rather than from the inside out. As such, we're still just beginning to approach the near experience of virtual worlds. We can still work to speak into being new and positive visions for the far experience of virtual society; to help build new worlds that augment rather than diminish our current one, worlds that reflect and magnify the best, rather than the worst, of humanity.

Let's return for a moment to *Ready Player One*. The film ends with protagonist Wade Watts, having won the treasure hunt and assumed control of the OASIS, deciding to shutter the virtual world for two days per week so that participants can reconnect with the real world. This ending is a fundamentally depressing one, insofar as it implies, among other things, that the OASIS was never built with its users' psychological fulfillment in mind; that it began with graphical immersion, only for things to go wrong from there. Real-world developers have been making this mistake since the earliest days of virtual reality. In order to avoid the dismal virtual futures envisioned by fiction, we must recalibrate our thinking and understand the sort of complexity that really matters—and has always mattered—in virtual environments.

A TALE OF TWO WORLDS

In 1990, the apparent future of digital computing came to a shopping mall in Chicago. Alongside a children's museum and a theme restaurant sat the multimillion-dollar BattleTech Center: the world's first consumer-facing application of a brand-new technology called virtual reality. For a mere eight dollars, shoppers and tourists could enter an enclosed "cockpit" and spend ten minutes playing what one newspaper called "the world's most sophisticated computer game." The cockpit was filled with a large monitor that displayed the game's relatively advanced graphics, and a microphone that let you talk with your fellow players in the other cockpits. "I really draw a distinction between the arcade business and this," BattleTech Center co-founder Jordan Weisman told a Chicago television reporter in 1990. "It's pretty much like comparing a merry-go-round to Disney World."

The excitement about the BattleTech Center typified the hype that surrounded virtual reality (VR) in the 1990s. The prospect of stepping into a crisp, realistic virtual world was irresistible for investors and technology commentators. Some even predicted that the virtual world would soon offer experiences more exciting and fulfilling than the ones available in our own. "If [people] could, say, play virtual basketball with a virtual Michael Jordan, then they wouldn't work, they wouldn't eat, they wouldn't bathe," one pundit told *Sports Illustrated* in 1991. "Next to VR, reality is just not what it's cracked up to be."

The pitch was convincing enough for plenty of tech investors, who sunk millions of dollars into virtual-reality research and development. The BattleTech Center soon expanded to a couple dozen shopping malls all over the world, and other consumer-facing VR applications quickly followed. The selling point, everyone thought, was the immersive, complex graphics that characterized virtual reality. Whereas with other computer games you stared at a screen when you played, virtual reality put you inside the game. Companies developed stereoscopic headsets that offered a 360-degree vantage into the world of the game being played, and digital gloves that turned the wearer's hand into a controller. Graphical immersion was what gave these virtual worlds their value, and what, to many people, made them feel like the future.

But the future never came to pass. By the end of the 1990s, many BattleTech Centers were closing and the virtual reality trend had subsided. The hype fell flat for several reasons. First, the concept of virtual reality had outpaced the era's available technology. Virtual reality headsets were bulky and uncomfortable, and far too expensive for at-home use. For all the promise and potential of VR's immersive graphics—and the graphics *were* advanced for their time—they still weren't objectively all that great, and they looked nothing like the real world. Creating an explorable photorealistic gaming environment is more difficult than creating a static photorealistic painting, in large part because graphic artists are limited by the capacity of the hardware and software tools available to them—and, in the 1990s, that capacity wasn't very high. It's hard to paint like Richard Estes if all you've got to work with is a box of crayons.

Even if you could have afforded your own VR headset, there wouldn't have been very much for you to do with it. In the 1990s, most internet users were still connecting to CompuServe and AOL on dial-up modems. High-speed broadband access was still largely the province of universities and large institutions. Without the ability to connect to a network of other VR users, you'd just end up wandering around a digital space by yourself, playing whatever games you'd purchased until you grew tired of them.

And people got bored with them pretty fast. Though the world of VR was graphically complex and visually immersive, it was experientially barren. VR games functioned along the lines of a trompe l'oeil painting, as if the graphics were backdrops that aimed to convey the visual illusion of depth, while offering nothing tangible with which the participants could meaningfully interact. You couldn't wander through these worlds and determine your own agenda. You couldn't meaningfully connect with other participants, because there weren't that many—the era's computing infrastructure simply didn't support lots of real-time simultaneous connections to such a resource-intensive environment, which in turn limited the value that could be created within these worlds. There just wasn't all that much to do in these virtual worlds other than to play the game, and even that experience had diminishing marginal returns.

The first wave of VR failed not just because the graphics weren't hyper-realistic or because the infrastructure wasn't yet in place for it to succeed, but because people couldn't find fulfillment within these virtual worlds. The takeaway, then as now, is that graphical immersion isn't enough to sustain a useful virtual world. Immersion without experience is just a visit to a wax museum filled with eerie, static figures, or, at best, a ride at Disney World: a rich environment, but one in which you nonetheless must stay on the preordained path. A virtual world that exclusively emphasizes immersion is one that calls out its own limitations, and in the process dissuades users from fully investing in its environment. Such a world is a product of insufficient complexity. While today's VR devices are far more sophisticated than their predecessors, many of the worlds they connect you to remain bland and empty. The energy spent on building better graphics still far outweighs the effort spent on creating sophisticated, living worlds.

Even back in 1990, perceptive observers realized that fulfillment mattered more than graphics when it came to virtual worlds. In a paper published that year, referring to the hype and investment being directed toward virtual reality, software developers Chip Morningstar and F. Randall Farmer wrote that "the almost mystical euphoria that currently seems to surround all this hardware is, in our opinion, both excessive and somewhat misplaced. We can't help having a nagging sense that it's all a bit of a distraction from the really pressing issues."

Morningstar and Farmer were also in the virtual worlds business. The pair co-developed a game called *Habitat,* which is arguably the first real embodied online virtual world. Launched by Lucasfilm Games in 1986, and run off an early online service called Quantum Link—a precursor to AOL—*Habitat* was an early experiment in radical online autonomy and individual need fulfillment. Players were free to explore the open world of the game at their own pace, according to their own interests and needs. They could choose their own avatars and create their own identities. If they wanted to go on a treasure hunt, they could; if they wanted to just stand around and converse with other players, they could do that, too. *Habitat* was a petri dish for self-determination.

Left alone to chart their own routes, participants went places that few would have predicted. One player created a newspaper within the world of the game, spending upward of twenty hours per week reporting, writing, and disseminating *Habitat*-centric news. Another player, a minister in real life, founded a virtual church, where he presided over virtual weddings between avatars. When, as was inevitable, some of those virtual unions ended in virtual divorce, the unhappy couples called on the services of *Habitat*'s in-world lawyers, who mediated the distribution of the aggrieved parties' virtual assets. Players even elected a virtual sheriff, who was charged with combatting virtual crime.

Though the game was meant to model the real world, *Habitat* did not look particularly realistic at all. It looked like a cartoon, and a rudimentary one at that. Avatars "spoke" to one another via speech bubbles. The backgrounds were blocky and lacked fine detail: Houses were big squares or rectangles, lawns were monochromatic strips of green. This comparative visual crudity did not bother Morningstar and Farmer, who believed that the inhabitants of virtual worlds had other priorities: namely, "the capabilities available to them, the characteristics of the other people they encounter there, and the ways these various participants can affect one another." As for the graphic sophistication of the technology itself, the two developers wrote, it was "a peripheral concern."

Habitat went offline in 1988, but the lessons of its open-world architecture and fulfillment-centric model have been reflected in a succession of massively multiplayer gaming environments that have redefined what it means to exist in a virtual space. From *Second Life* to *Minecraft* to *Eve Online,* countless games and virtual worlds have since presented players with open realms in which they are free, within the parameters of both the game and the technology that powers it, to choose their own destinies. These virtual worlds and others have chosen to emphasize complex interactions and player autonomy more than just their impressive graphics. In the process, they have become more and more integral to their participants' daily lives.

The massively multiplayer game *Eve Online,* for instance, has been an active virtual world since its launch in 2003. Nominally a role-playing space adventure in which players assume various positions as colonists of a new galaxy, the world and the experiences therein have proven so consequential to participants that one man, former journalist Andrew Groen, has taken it upon himself to act as *Eve Online*'s unofficial historian. Groen has published two hardbound books recounting the in-world history of the game from 2003 to 2016: wars fought, alliances made, memorable personalities and significant events and other things of note, all reported and fact-checked and set within a clear historical chronology. The books are not histories of how the game was developed: They are histories of things that happened within the world of the game. Their very existence is a testament to the centrality of the *Eve Online* world to so many people's lives.

More than thirty years after *Habitat* and *BattleTech* launched, digital environments have become more fulfilling, useful, and ubiquitous than ever before. Meanwhile, the visual experience provided by virtual reality has gotten a lot better, as have the graphics available on the leading edge of modern gaming. But even though graphics technology is better than it's ever been, many of the most important games of our own era are actually of lower graphical fidelity than they otherwise might be. Take *Minecraft,* for instance, the wildly popular open-world game with a functionally infinite amount of terrain for its players to explore. Though *Minecraft* is a complex world, its graphics are deliberately crude. The avatars and terrain are blocky and cartoonish. The game's visual aesthetic is many levels below the optimal visual experience that today's best technology can deliver.

Its players don't seem to care that much: As of August 2021, *Minecraft* boasted more than 141 million monthly active users. People don't get sick of *Minecraft.* People keep playing the game, and they keep finding ways to make its world work for them, rather than the other way around. While the nominal objective of *Minecraft* is to mine materials that can then be used to craft various items, in practice participants can choose their own adventures within the game's virtual world. For example, many kids will log on to *Minecraft* when they get home from school in order to visit with their friends from school. The game has become something of a town square for the younger generation. *Minecraft* is a conduit for useful, fulfilling experiences—even if those experiences aren't particularly immersive ones.

Ideally, of course, we'd want both immersion and experience within our virtual worlds—top-tier graphics and top-tier interactions—and the good thing is that, technologically and culturally, we're approaching that point. Today's digital games can provide their players with actual interactive worlds, rather than just compelling visual illusions. Meanwhile, improvements in computer processing power have allowed us to render and present these worlds—and the opportunities available within— to thousands of people at once, in something approximating real time. Finally, both high-speed internet access and networked devices are now widely available. Decades after first hearing that we'd all soon be shooting hoops with a virtual Michael Jordan, or traipsing across virtual terrain in search of virtual adventure, the metaverse is finally catching up to our ambitions.

As an entrepreneur working in this space, I am now and then privy to demos and experiences that offer jaw-dropping glimpses at what might be possible in the metaversal future. In May 2021, for example, my company, Improbable, ran a demo that packed 4,144 separate avatars, controlled by 4,144 separate people, into the same virtual space all at once. The goal of the demo was to make it so that every human-controlled avatar could see, hear, and react to every other human-controlled avatar within the world: to create intimacy at scale.

There was indeed something profoundly intimate yet vast about the demo, about having so many autonomous avatars all existing and working together at once in a visually dense and immersive virtual space. There was a collective effervescence to the experience that felt like a teaser of what's soon to come. At one point in our testing, we enabled large-scale voice, and a horde of hundreds of strangers began simultaneously singing "Africa" by Toto. An outpouring of emotion accompanied this event, where the metaverse suddenly seemed to come to life. One's mind abruptly shifted in its acceptance of that reality. You felt weirdly self-conscious, knowing that so many people could hear your voice. The virtual worlds of the near future will be intimate at an even more massive scale. They will combine the open-world architecture of *Habitat* with the immersive graphics long promised by virtual reality in order to provide useful experiences that will satisfy both the senses and the soul. Let's discuss how to measure the complexity we'll need in order to make this happen.

USEFUL COMPLEXITY

We've long known how to quantify and assess visual immersion in digital environments. We do so by considering both graphics—pixels, the refresh rate, resolution—and the extent to which the environment follows the same natural laws by which the physical universe is bounded. From those inputs we can empirically understand the variance between what a computer can present and what we can see in the real world with our eyes, then judge how successful the computer was at creating a convincingly immersive space.

The early days of computer games were defined by experiences with levels of graphical fidelity that were only a fraction of what humans would need in order for the experiences to actually look real. Conversely, the graphical experiences offered by the most prominent fictional virtual worlds generally feature extremely high immersion levels. It's not hard to understand how a developer might use immersion metrics to assess the gap between the simulation and the actual, and to try to iterate a more immersive product in subsequent versions of the game.

In order to approach peak usefulness, we would need similar methods for quantifying and thus evaluating the experiences and fulfillment that a world can provide. A virtual world is a place that generates meaning for its inhabitants. How do we assess its ability to provide that meaning not just for a single user, but for every simultaneous participant in that world?

There are two main criteria by which we can evaluate experiential complexity in virtual worlds: how rich the world is in terms of individual interaction, and how well that world can support the vast simultaneous web of changes needed to sustain a society. Taken together, these criteria combine to measure a given world's *useful complexity*. The term in this context means a few things. For one, it refers to the world itself, and the quantity and quality of the possible experiences that can be had therein. A virtual world's useful complexity increases with every new object that a participant can pick up, hold, and deploy; every new environment that a participant can explore; every new avatar with whom one can meaningfully interact. The baseline here—the thing against which the utility of a virtual world must be measured—is the real world. The real world is nothing but depth, and the fact that we tend to take this depth for granted is a testament to its reliability and seamlessness.

Imagine yourself at a house party, in a room filled with people. Though you might not immediately realize it, there's a functionally infinite number of things you can do there. You can interact with every single item in the party room: You can sit on the chairs, flip the light switches, browse the books on the host's bookshelves, paw your way through a bowl of popcorn, grab a bottle opener and use it to open a beer. You can work your way through the room, interacting with every single person in it, talking and laughing with each of them; you can also interact with them nonverbally, nodding and waving to people whom you like, or studiously trying to avoid those whom you don't. You can rip off your shirt and put a lampshade over your head in a bid to become the life of the party. You can hug people. You can physically fight them. You can steal away and roam through the rest of the house in order to get some peace. There are innumerable opportunities for you to interact with the party environment and everything in it—and depending on how those interactions play out, they might carry short- or long-term consequences for the future. (If you pick a fight at a real-world party, for instance, you might not get invited to the next party.)

Even a familiar real-world environment can be rich with interactive possibilities. And here's another important thing to note: Everyone else at the real-world party is also having their own individual experience of the party at the same time you're having your experience of the party. (The real world is great at providing intimacy at scale.) No matter how many people cram into that room, their individual experiences of the party will be limited primarily by their own choices, not by the environment itself. Yes, at a real-world house party, it gets warmer inside the house and it takes more time to get to the kitchen as more and more people arrive, but the house itself doesn't glitch and take longer to render when it is stuffed with people.

Crucially, all of these individual experiences combine together. Every participant can simultaneously experience every change in the environment, which can create wonderful and profound opportunities for emergent experience. We can intuitively understand how the connectivity generated when lots of people come together can create opportunities that would be otherwise impossible. Think of the magic of a spontaneous chant in a football stadium, for example, one that requires rapid coordination among thousands of people reacting to one another at once.

Fictional representations of virtual worlds, such as the OASIS in *Ready Player One,* are able to seamlessly present all of their participants with infinite and equivalent depth. For technical reasons, though, today's virtual worlds have heretofore struggled to match this sort of useful complexity. It takes an immense amount of processing power to create and render a digital environment in which every single object or individual is equally interactive, and which can sustain an ever-increasing number of participants while providing the same experiential opportunities for all of them. But, even more important, this problem has absolutely nothing to do with building better graphics. It's a problem of communication and networking.

A system must be able to understand and apprehend all the different connected participants, and to juggle their access to information—sort of like a vast air traffic control tower, albeit one that must operate at unfathomable speed. While this book is not meant to focus on the computer science of the metaverse, it's worth briefly highlighting that this problem is much harder to solve than it seems. Even huge modern systems like Google Search or Amazon's store are based on solving what are known as "embarrassingly parallel" problems. When two people both make a search request or try to buy a pair of socks, there is no need to exchange information between these participants. These requests can just be shuttled to different processes to handle. The problem of building a dense interactive social reality, however, cannot be solved this way, since building the perspective of one user requires some knowledge of what every other user is doing, in real time. As the number of users grows, the problem becomes quadratically harder.

In the next section, I'll talk more about exactly what it takes to maximize these sorts of communications operations; for now, though, suffice it to say that with every step we take toward making virtual worlds more universally and simultaneously interactive, we also get closer to optimizing the useful experiences that can be found therein, and the extent of the societies that these worlds can support. The day will eventually come when a party in a virtual world feels even more true to life than a party in the real world, because there will be more useful experiences available in a metaversal party than in its real-life equivalent. You might be able to perceive it from many perspectives at once, for example; you might be able to scan a crowded room and immediately know every partygoer's name and occupation. For now, though, we're working on getting close to the real world's level of depth, one virtual party hat at a time.

The good news is that a virtual world needn't exactly mirror the real world's level of depth in order to support a society. *Habitat,* after all, was a cartoonish environment that nevertheless supported an experientially complex society. Virtual societies can and will arise whenever a world is sufficiently capacious, resilient, persistent, and consequential. Complexity emerges naturally from open worlds that facilitate user interactions, present a variety of valuable experiences, and allow users to chart their own courses within the world. These sorts of worlds are ones in which complexity can also be deployed to solve real-world problems.

If a world is complex enough to accurately model real-world scenarios, then we can begin to use it to support optimal methods of rehearsal for all sorts of jobs and pastimes: urban planning, disaster management, product rollout and development. Eventually, virtual worlds will function as "what-if machines" that we can use to answer questions about things, so that we can go from a society that is often stymied by complex problems to one that can learn how to better manage complexity in the real world. (I'll note that this vision will require vast increases in the computational complexity of such virtual worlds.) These worlds will be useful on both an individual basis and a social basis. They will create real, lasting value for their users and for society.

The goal for developers must be to create environments that are useful in all of the ways mentioned above: ones that can sustain countless individual simultaneous interactions with countless individual items and users, while providing escalating opportunities for psychological fulfillment, and also creating value that can help to make the real world run more efficiently, effectively, and intelligently. When a virtual world can consistently fulfill these three needs, we'll know we're approaching true depth and maximum usefulness.

But what will be our yardstick? How can we quantify the progression of the journey toward maximum usefulness? A lot of the talk around virtual worlds lacks empirical rigor—I'll have more to say on this point in the next chapter—which leads to a lot of unrealistic promises and expectations, and which keeps the popular image of the metaverse tethered to its fictional representations, and to the blustering of its most self-interested promoters. A viable metric for usefulness will make it easier for neutral observers to discern hype from reality, and it will give developers something concrete to strive for as they work to build virtual worlds that, at their peak potential, may well turn out to be more useful than our own. The metric that I consider best was devised by my co-founders at Improbable and is called *communications operations per second*. Think of it as the megahertz of the metaverse.

THE MEGAHERTZ OF THE METAVERSE

If you were tasked with devising a single video game that a toddler, an orangutan, and a Martian could each equally well understand, you could hardly do better than to just replicate *Pong*. Released by Atari in 1972, the first commercially successful video game remains one of the simplest in the history of the medium. A simulated version of table tennis, *Pong* features two "paddles"—thick white lines that move on a vertical axis at either side of the screen—which players use to volley a tiny ball back and forth over a dotted-line "net" that bisects the screen. If your opponent fails to return your volley, you score a point. For all intents and purposes, that's the game.

There are three moving parts in a game of *Pong*: the two paddles and the ball. At any given moment, there can be at most three separate and simultaneous things happening. Operations per second is a measure of how many separate and simultaneous things can happen in a virtual environment, by reflecting how many messages are being sent or must be sent simultaneously to model that environment. As an example, at the time of writing, a game of *Fortnite* that allows 100 players to interact together requires roughly 10,000 communications operations per second. This statistic means that the server needs to process all of these messages, and also to quickly send them to the machines of each connected user that needs them.

Adding more players, and having to synchronize more and more information, would eventually impose such a burden on the server that the game would slow to a crawl and then crash. The more interactive the game becomes, the more operations per second are required, since more information needs to be synchronized. If suddenly the world is full of angry tigers, then all those tigers will create changes that must be propagated to every other participant who can see them, creating an ever larger communications burden. The more operations per second that the world is able to support, the more rich, realistic, and immersive that world becomes. The metric quantifies the horsepower that, at its upper limits, can turn virtual reality into something approximating or exceeding actual reality.

To judge by the seamlessness of the interactions that occur there, most of the virtual worlds that we see in fiction boast a functionally infinite level of operations per second. But infinity is not achievable in the real world. Rather, our maximum operations per second are limited by the capacities of the technologies that undergird the virtual worlds we build—and as we approach those limits, our virtual worlds can start to feel a bit shaky. Imagine that someone has placed a one-kilogram weight on your shoulders, and that every few seconds, that weight increases, one kilogram at a time. You'll be easily able to bear up under the weight for the first few minutes, but there will eventually come a point where you will start to struggle. Your knees and back will shake; you will start to sweat and strain. Eventually, no matter how hard you try to stay upright, the weight will become too much for you to handle, and you'll collapse under it. No matter how strong you are, there's always a collapse point.

The same has always been true for virtual worlds and the operations-per-second levels they are able to sustain. It would take a staggering number of operations per second to convincingly simulate the real world. Let's return to the previous example of the virtual party as a means of illuminating this concept. A maximally useful virtual world wouldn't just have to be able to sustain and present the countless simultaneous communications operations happening at one virtual party—it'd have to be able to do so for hundreds of virtual parties happening simultaneously all over its world. It'd have to be able to sustain a virtual concert, where 50,000 people are together in the same space. It'd have to be capacious enough to let us simulate an entire war.

At that point, we'd be looking at billions, even eventually trillions, of communications operations per second, served to connected clients all over the world, at low enough latency to support real-time interactions. This challenge is an absolutely colossal one, and it is often conveniently hand-waved away when companies release slick trailers that promise such futures. Even if you're able to support such scale, you would then encounter new, related challenges. How on Earth would you test such infrastructure? How would you prevent it from falling victim to hackers and exploits, or get it to operate with sufficient stability to be entrusted to manage high-value digital assets? Companies such as Google have built modern miracles with their search infrastructure, which far outstrips the complexity of almost anything else on the planet, but even this type of capability is not even close to what would be required to build a true representation of the real world that everyone could enter—not by orders of magnitude.

As a rule of thumb, when building difficult distributed systems—types of computer systems in which many different devices must interact in order to run some capability—as scale increases by a factor of ten, you need completely different types of architecture to manage the growth and the ever-harder challenges that it poses. Today there are lots of companies claiming to be a hop, a skip, and a jump away from the metaverse—and yet they've offered almost no demonstrations of actual working solutions to the complex problems described above. I hope that this chapter has armed you to question the utility of shiny graphical demos as proof that a product or business can actually support useful virtual worlds.

We will eventually get to the point where virtual worlds are supremely useful—and, as indicated above, the qualifying factor will be a matter of scale. As the number of people simultaneously engaged in a virtual world grows, so too does the value of the world. If millions of people are engaged in a virtual world, and all of them are having rich, complex, fulfilling experiences, then that world will be immensely valuable. Soon it will begin to generate not just intrinsic worth, but significant extrinsic worth, too. We will then be able to quantify the value that the world presents, and that world will start to become an economic force. Once we get worlds that are powerful enough to optimize experiences and adjust for peak fulfillment and utility, we'll achieve the sort of VR experience promised thirty years ago— the sort that you don't want to leave not because you're trying to escape your "real" life, but because the virtual world is a seamless extension of and improvement on your real life.

The many dystopian visions of virtual worlds can be said to have gotten at least one thing right: the idea that the real world is indeed verging on chaos. One salient question as we approach the era of virtual society is whether we will use our digital worlds to escape a struggling world or to help set it right. In science fiction, that question has already been answered. Fiction tells us that when we flock to virtual worlds, we simultaneously abandon our own. In the real world, though, we can make different and better choices. We can take the lessons from virtual worlds and use them to help fix many of the things that are currently wrong with the real world, from social, economic, and political perspectives.

But virtual worlds will not meaningfully affect the real world as long as they stand apart from it. We must forge an indelible connection between virtual worlds and the real world if the value created in these digital realms is to be transferable into our own. This nexus between worlds is what we mean when we talk about the metaverse. Over the next few chapters of this book, I will dig into the details of how we might build and sustain a valuable one.

Chapter 5

A NETWORK OF MEANING

In June 2021, Facebook CEO Mark Zuckerberg announced that he was staking the future of his company on the metaverse. Zuckerberg proclaimed that the social-media giant would create an immersive virtual world which would transform work, play, gaming, shopping, and, well, life as we know it. The metaverse was the future of the future, and Facebook—which would soon rename itself "Meta"—was planning to build it.

The impact of this announcement, which was clearly also an attempt to distract public attention from Facebook's other problems, was dampened a bit by the inescapable sense that neither Zuckerberg nor anyone else weighing in on Facebook's announcement could actually define what the metaverse was or would be. Vague allusions were made to virtual reality and avatars. The metaverse was presented as a space without limits.

None of this rhetoric was particularly helpful in explaining what, precisely, the metaverse is, or why it would be worth using. Facebook's vision of the metaverse amounted to vaporware: a theoretical space in which users could do almost anything, and thus an idea that, practically, amounted to nothing.

In the following months, as Facebook worked to remake itself around the metaverse, countless others frantically followed suit, working overtime to establish their own visions for the metaverse and precisely where their companies and products would fit within it. The explosion of interest and activity around the metaverse did not serve to clarify the concept. Quite the opposite. More than a year later, with billions of dollars and countless work-hours having been spent on trying to speak the metaverse into being, it still sometimes feels like the discourse has been actively evading many of the foundational questions that must be answered if digital metaverses are to be worth a damn to anyone other than the oligarchs who hope to build and own them.

I contend that you can't begin to understand the metaverse without first understanding the *how* and *why* of the virtual worlds that will exist within it, and that have existed throughout human history in other forms. This book up until now has been primarily concerned with laying the groundwork for this understanding. We've explored how virtual worlds, which have been a focus of human imagination and ingenuity for millennia, will in their modern forms offer unprecedented access to a wide variety of useful and enriching experiences, and how participating in these experiences will in turn improve people's lives. As I wrote in Chapter 3, technology now allows us to create, refine, disseminate, and evolve useful experiences with unprecedented depth, breadth, speed, and precision. Eventually, we'll have virtual spaces that can support millions of simultaneous participants, all of whom will have access to any sort of digital experience that could possibly exist. These spaces will thus become powerful vehicles for human fulfillment, and they will create psychological, social, and economic value on an unprecedented scale. We have not just the opportunity but the obligation to build them.

A slick corporate metaverse compelled by glib visions of marketing synergies won't be a valuable metaverse at all; it would, in fact, be fundamentally at odds with how a metaverse actually creates value. With profit, central control, and opacity as its orienting philosophies, it will divide people rather than unite them. But even if the Facebookverse takes shape, and even if the company's vast resources give it advantages over others, there's no reason to think that bland corporate versions of the metaverse will be the only versions. Just as there's more than one website, more than one movie, more than one video game, and more than one telephone service provider, there will be more than one digital metaverse—many more. There have been a multiplicity of metaverses throughout history, and the same will be true in the digital sphere. There will probably also be a metaverse of metaverses—a *mega*verse, perhaps—that connects the various metaverses together. (The M^2 project—which, full disclosure, my company built—is one of the first attempts to create an internet of metaverses.) Some of these metaverses will be more prominent than others, and some will be more valuable than others. But where will this value reside?

With the groundwork about virtual worlds and human psychology now laid, it's finally time for us to return to the task of defining the metaverse, as we approach its proverbial front door. Will the metaverse be a centralized virtual space with multiple points of entry, or merely a loosely related constellation of virtual world–type experiences? Are the sorts of game-like virtual worlds that first jump to many people's minds when they think about the metaverse all that there is to the concept? In this chapter, I'll take a close look at some of the many extant definitions of the metaverse in order to show how vague terminology serves to inhibit understanding. I'll explain why it's so important to have a cogent working definition for the metaverse, and why loose definitions are both the products and the progenitors of lazy thinking. I'll draw on historical examples to explain why bilaterality between worlds is a salient characteristic of a functional metaverse; from there, I'll explore the ways in which a metaverse is a network of meaning and consequences, the constituent parts of which grow and expand via the ingenuity of its participants.

At the dawn of the internet, society did not yet fully understand where the value in the network would lie. The choices made by investors and developers between then and now—choices born out of definitional laxity and general short-sightedness—have made it much more challenging to tame the sorts of network excesses for which social media platforms in particular have shown themselves to be primary vectors. The metaverse offers an opportunity to improve upon the internet and avoid the mistakes we made with that medium, so that we can work to create new meaning out of megahertz. But if we are to separate the hype from the substance and set our expectations accordingly, then first we must understand what, exactly, the metaverse is—and what it isn't.

METAVERSAL ONTOLOGIES

Ask ten different people to define the digital metaverse, and you're liable to receive ten very different answers—answers which, taken together, promise everything and tell us next to nothing. This broadness isn't nefarious in nature. *What is the metaverse?* is still a legitimately hard question to answer. While proto-metaverses—virtual worlds such as *Minecraft* that have developed from gaming platforms, and have made some small mark on real-world society—currently exist, a metaverse with the kind of utility envisioned by this book does not yet exist, and all of the people who are working to visualize and build it are working from slightly different assumptions and expectations.

The metaverse has been defined variously as an "even more immersive and embodied internet" (Mark Zuckerberg), a "living multiverse of worlds" (Jon Radoff), a "massively scaled and interoperable network of real-time rendered 3D virtual worlds which can be experienced synchronously and persistently by an effectively unlimited number of users with an individual sense of presence, and with continuity of data" (Matthew Ball), an "engaging digital landscape" where you can "bicycle, surf, motorcycle, drive, compete, tell stories, be told stories" (Strauss Zelnick), and "an aspirational term for a future digital world that feels more tangibly connected to our real lives and bodies" (*The Verge*), to cite just a handful of the more cogent definitions I could find. There are plenty of others that I won't bother citing here.

It's worth noting that the confusion over the metaverse presents more than just an aesthetic problem. First and foremost, these definitions are inconsistent, and inconsistency leads to strange behaviors from investors and creators. Should the billions of dollars currently pouring into metaverse projects go to virtual reality and immersion, or somewhere else? Will the killer app of virtual worlds be the ability to tell stories while motorcycling through a living multiverse of worlds, like some futuristic Che Guevara? This ambiguity means that lots of people are going to throw good money after terrible ideas.

If you're planning to invest your money, time, or attention into a new, emerging product or project, then it would be wise to understand what the product is and how it creates value. If your investment is guided by a weak or inaccurate definition, then the results can be dismal for all concerned. In 1996, for example, the fast-food restaurant McDonald's introduced an upscale hamburger called the Arch Deluxe, hoping to appeal to the adult palate. The Arch Deluxe featured "gourmet" ingredients and was sold at a price point to match. McDonald's spent at least $200 million launching the sandwich, which today is remembered as one of the biggest marketing disasters in restaurant history. As it turned out, nobody wanted "upscale and expensive" from McDonald's; they wanted to spend forty-nine cents on a burger they could eat in their car. McDonald's lost sight of what sort of restaurant it was and what people wanted from it, and the company lost lots of money as a result.

Let's pivot to a more directly relevant example. Around the turn of the twenty-first century, there was widespread confusion over the core function of the internet and the World Wide Web. While a lot of people were excited about the Web, much of that excitement was fueled by various inaccurate or premature claims about what it was for—virtual currency! one-hour DVD delivery!—and the ways in which people would use it. Consequently, lots of investors made big bets on websites that promised more than they could deliver, and many of these investments failed spectacularly. The business ideas that did pay off were those that worked within the limits of the early internet—such as selling books online, for instance. While it was expensive and difficult to use a computer to buy something online back then, the experience was worth the hassle if the online store had a significantly better selection than a store at the mall. Books fit perfectly into that niche. The source of value for many internet businesses turned out to be connecting buyers and sellers more efficiently, something not as obvious back then as it seems now.

A valuable definition of a product or a service must illuminate its core functionality, especially when individual designs and implementations will vary wildly. When talking about the metaverse, it's irresponsible to just wave our hands willy-nilly and say that video games are a metaverse, digital worlds are a metaverse, Disney Plus is a metaverse, everything's a metaverse. These expansive conceptions of the metaverse work only to the advantage of some of the people who are eager to hype it. In the absence of definitional rigor and a common frame of reference, any prediction or promise can seem equally plausible—which can make charlatans seem like savants.

The rest of us are ill-served by this chimerical cycle. Ontological confusion can breed fear, contempt, and cynicism among people who are being told their lives will be changed by a thing that they have no clear way to understand. It can also cause people to make losing investments, and, once burned, to back away from the sector entirely. So let's be meticulous when working to define the metaverse, and let's proceed with clarity and intent.

Where do the existing conceptions of the metaverse fall short? When I sat down to write this chapter, many people were stuck on the notion that a metaverse is an "even more immersive and embodied internet," as would-be metaverse maestro Mark Zuckerberg has put it. This definition is not a particularly useful one. Even seemingly wiser definitions, such as Matthew Ball's contention that the metaverse is "a massively scaled and interoperable network of real-time rendered 3D virtual worlds which can be experienced synchronously and persistently by an effectively unlimited number of users with an individual sense of presence, and with continuity of data," aren't all that useful when you dig deeper. For one thing, these definitions are very broad. There are already lots of products and applications that represent virtual worlds in various forms; from *Habitat* to *Ultima Online* to *Second Life* to *Minecraft*, we've been working with and within virtual worlds for over thirty years. Saying that the metaverse is just another virtual world—or even just a massively scaled network of virtual worlds—renders the concept merely a buzzword for more advanced video games.

"Virtual worlds, but better ones" is a good idea—previous chapters have explained *why* it's such a good idea—and we should definitely try to build them. But that definition is insufficient, too. If a metaverse is, at core, a virtual world, then how do we measure its value? Is the value of a metaverse simply tied to the fact that it is a virtual space that doesn't physically exist? If so, that's a pretty low bar. Or perhaps the value is related to the world's immersiveness and embodiment? If that's the case, though, then it would follow that the virtual-reality worlds of the 1990s were more valuable than the early open-world games of the same era, and we've already shown the deficits of that particular comparison.

Let's try again. Maybe it's not enough to say that a metaverse is just a virtual world. Maybe a metaverse must be a really, really advanced and immersive virtual world, way beyond anything we've already built—something like the digital world of *The Matrix*, or the Holodeck from *Star Trek*. Both the Matrix and the Holodeck are the sorts of virtual worlds that virtual-reality enthusiasts imagine when they envision VR technology at its apex. They are embodied, three-dimensional, open-world spaces, and, in the case of the Holodeck, people go there explicitly to find fulfillment. It would be a true technical accomplishment to build a virtual world that was as expansive and seamless as the Matrix or the Holodeck. But would that world also qualify as a metaverse?

I would suggest that "a really great virtual world" isn't a useful definition for a metaverse, either. For one thing, if a metaverse is just a complicated virtual world, then that wouldn't explain why there's such a powerful cultural rush for so many real-world products and entities to get involved with these virtual worlds. I don't remember any *Star Trek: The Next Generation* episodes where Riker and Picard loaded up a Holodeck adventure in which they toured the factory where smooth, refreshing Earl Grey tea was made, after all. The Holodeck and the Matrix stood apart from the real world; they weren't integrated with it in the way that tomorrow's metaverses clearly will be. The Matrix, you will remember, was born out of a machine dystopia; it was integrated with the real world in a way that brought no value to the world's human inhabitants.

Even if we accept the definition that a metaverse is just a really great virtual world, we would still run into some comparison problems. Is the Matrix a better Holodeck? After all, they're both equally immersive. Is the Matrix more immersive because you connect to it by putting a thing in your head? Is the Holodeck more immersive because you enter it by choice, rather than being forced into it by malicious machines? How can we empirically evaluate which world is more successful? We can't do this on anything more than a very superficial level, which tells us that immersion alone isn't the best way to define a metaverse. Unfortunately, the majority of metaverse companies currently out there are focusing on immersion as their key measure. They believe there is a direct correlation between how good the graphics are or how real the world seems and how valuable the metaverse might be. This line of thinking is flawed, and it directly wastes investment capital.

Perhaps the value in a metaverse can be measured by assessing the feelings of immersion, autonomy, competence, and fulfillment that it affords its users. Given that we've already covered why these are good ways to think about the value of experiences, it's perhaps not a bad starting point—but by this definitional logic, a dream would qualify as a metaverse, which seems like a problematic conclusion. After all, a dream is a totally immersive, totally present space in which participants can have many useful, fulfilling experiences. Dreams are also capable of changing people's lives, as shown in the lucid dreaming research of Stephen LaBerge. A good dream can be configured for optimal fulfillment, insofar as the limits and laws of physics, nature, and economics do not generally apply; the dreamer is generally the main character in their own story; and the dream itself generally revolves around the dreamer. So is a dream the ultimate metaverse? If so, then why wouldn't we want to dream all our lives?

We already know that this logic is flawed. If spending life in a dream state was the best way to extract the maximum value from your time on earth, then no one would ever wake up. People would seek out sedation, and *Sleeping Beauty* would be recategorized as a self-help book. And yet we know that sleeping forever would be a bad thing, as opposed to a desirable outcome, because we intrinsically understand that there is great value in being able to wake up and exit your dream. Dreaming our lives away would alienate us from society and one another. Participating in society and existing alongside our fellow humans is important, because, ultimately, people do not seek isolated fulfillment. Much of what we depend on to live meaningful, fulfilling lives involves other people, social realities, and a world of consequences for our actions and choices.

The ability to relate to one another within the context of society is fundamental both to our existence as a species and to the value that virtual worlds have created for the real world for millennia. Let me now start to propose the basis of a better definition of the metaverse. If a video game is something that an individual plays and derives fulfillment from, then a virtual world within a metaverse is effectively a game that *society* plays together, and a metaverse is the structure that mediates the transfer of the value created by engagement in that virtual world back to the real world and between virtual worlds. The absence of these social factors and of any conduit for value yields the horror found in the morality-play virtual worlds of science fiction—escapist fantasies that serve to distract and deflect from the outside world, in which no value from the other world flows back to the real world.

Popular definitions for a metaverse are often incomplete or deficient because they are either synonymous with existing concepts, lack the precision necessary to help us understand and evaluate a metaverse's core function, or describe an environment that, when gamed out to its natural end points, would end up alienating its participants from the societal context in which they lived. In order to arrive at a more useful definition of a metaverse, we must explore how those virtual worlds interact with one another and with the real world. We must examine the relationship between the social realities in which people live on a day-to-day basis and the constructed realities in which people choose to invest their time, attention, emotional energy, and ingenuity.

MEANING AND THE METAVERSE

As you'll recall from Chapter 1, we have always had the capacity as a society to believe in, and to imbue with a kind of half-life, worlds of events, ideas, and people that are not strictly real. We have engaged this capacity for millennia. The primary function of these worlds is neither explanatory—that is, they do not mainly serve as stories we tell in order to explain how the real world works—nor escapist. Instead, these worlds actually become the basis for us to have activities that we find fulfilling. Society is compelled to play the games these worlds represent, and it needs to make these worlds feel socially real. These worlds create lasting meaning for society not just because of the fulfillment that individuals derive from their participation in them, but through the ways in which these virtual worlds affect, intersect with, and create value within the real world.

From the Egyptian cult of the dead that spurred the construction of the pyramids to professional sporting fandoms so intense that they can spark riots after negative outcomes and parades after positive ones, imagined universes throughout history have existed in direct conversation with and relation to the real world. This bilaterality—the ways in which each world mutually affects the other—is what makes them metaverses, as opposed to just virtual worlds or engrossing stories.

The people who suggest that a metaverse is simply a rich virtual world with various rich experiences therein are missing the point. A metaverse is more accurately described as an "other world" of living ideas that intersects with our own world in various ways. These worlds of ideas feature shared histories, shared economies, and imagined sets of events or states of affairs that serve as the basis for a mythology. They are populated by personalities, events, and things whose persistence is powered by the collective belief in their existence. Those things in turn are connected to and have real consequences for the society that creates them.

Even so, the concurrent existence of two or more worlds isn't enough to qualify as a metaverse, either. If intelligent life were to be discovered on Mars and Venus and we were to establish interplanetary trade routes and settlements there, that wouldn't make Mars, Venus, and Earth a metaverse—that's just a rather lively solar system. The other worlds I've been writing about are real in a very careful, socially constructed way, and the consequences of these worlds' realities with respect to your "home reality" are also carefully moderated.

For example, many ancient metaverses were characterized by the idea that when you died, you went to that other world, and that your words and deeds in this world in some way affected your placement in the next. The other world was a real place, insofar as people believed that you actually went there when you died, and that you couldn't come back. These rules of transit indicate that while there is a relationship between the worlds in a metaverse, there is not generally a direct convergence between these worlds. Value is created and exchanged between worlds at the point of contact between the virtual world and the real world. The social construction of these other worlds allows them to create opportunities for their adherents, instead of just creating further problems.

The defining characteristic of a metaverse, then, is the way in which it generates a network of meaning and value between the real world and the half-life world or worlds that are all linked. A metaverse is a network of consequence and meaning, and participating in these networks allows us to become what I term our full metaversal selves. Meaning flows directly from the other world to the real world—and, in turn, back from the real world to the other world. The value derived from society's ongoing belief in the existence and merit of the other world manifests tangibly in the real world. The reach and richness of these other worlds can touch all facets of real-world society. They create cultural value for the real world in the works of art, music, literature, and architecture that they inspire. Belief in these other worlds can generate cultural traditions that unite and organize the societies that follow them; it can imbue otherwise banal events and daily activities with depth and resonance. These other worlds can inspire codes of law and behavior around which a society can organize itself.

If you accept that these other worlds of ideas can and do create tangible value in the real world, then it's easier to understand why we would want to progress from the ancient metaverses model or the sporting metaverses model to the "metaverse mediated by digital experiences" model. For all their utility in their time, ancient metaverses were fairly static in terms of the information and experiences they contained. Access to this information and these experiences, moreover, was usually mediated by a caste of priests and soothsayers who could expand, contract, explain, or abolish facets of the metaverse seemingly at whim. (Not so different from team owners and league commissioners in professional sports.) Consider the ecumenical councils of Nicaea and Constantinople in the fourth century A.D., for example, where the era's Christian leaders convened to, among other things, resolve schisms over the divinity of Christ and hammer out the doctrine of the Holy Trinity. These doctrines that today's believers take for granted were effectively the product of early Christendom's top clerics getting together and deciding what the faithful would all agree to believe.

Digital virtual worlds are much more directly and clearly

linked to the real world and open to input from ordinary people. Because they will offer more points of entry for meaningful participation for more and more people, digital worlds will in turn produce dramatically more economic and social value than their predecessors. It's not just that digital metaverses will feel more real and immediate than their historical antecedents—it's that they will facilitate more conversation between worlds, and will allow for more expressions of ingenuity from individuals. This model of metaverses is doubly powerful because it provides a new lens on the utility of cultural other worlds, beyond the role they play for the individual. Choosing to see engagement in other worlds such as sport or religion as a sort of productive game that society is playing offers a new way to understand the social utility of these metaversal pastimes.

A metaverse is a network of consequences and meaning between multiple worlds in which people are simultaneously engaged and invested. One or more of these worlds is made of ideas, and the other one is the physical world. This network is moderated through human-made rules, as opposed to natural or physical rules. When you die in *Fortnite*, for instance, you do not also die in the real world. Human ingenuity creates, sets, and expands the parameters by which the other world exists, and by which it relates to the real world. The term *ingenuity* is here used to refer to the skill of causing useful change in a metaverse by obeying and extending the social reality of the worlds therein. Ingenuity isn't the same thing as pure creativity. When you propose changes to a world within a metaverse, your changes must fit within the rules society has created in that other reality. The network of meaning within a metaverse is continually modified by a collective co-creation mediated by these explicit and implicit rules.

So, you have the real world. You also have another world, or many other worlds, that have been created by people in the real world. When you have fulfilling experiences in these virtual worlds, that sense of fulfillment remains even when you exit the virtual world and return to your life in the real world. This sustained sense of fulfillment is the simplest form of value transfer between worlds. It is also possible to create more tangible forms of value in a virtual world—such as fame or wealth—and to transfer those assets into the real world. A community might embrace rituals performed in some virtual world to the point where they might transform the context of their real lives. If you've ever become obsessed with an online game, you can understand the roots of this process.

There are many pathways by which value can flow through and between these worlds and to the real world. When you consider how much of our economy is already based on intangible value, you can start to understand just how important this transfer of value could become. The value being created and transferred can be social value, or intrinsic meaning, or a sense of identity, or a family of values. As these other worlds grow and become more complex, and as more and more people begin to contribute to them, the types of value that emerge from these worlds will also start to expand. The digital asset economy turns this value transfer into something even more tangible. The magic sword you acquired in virtual space is now as real as your stock in some company.

Metcalfe's law, which holds that the value of a network grows in direct proportion to the number of connected nodes within the network, applies to a metaverse both as the number of connected participants grows and as the number of useful experiences available therein grows. But the applicability of Metcalfe's law is based on how usefully connected the various worlds are. As a metaverse becomes more useful and meaningful, it also becomes more valuable. If the metaverse is just a bunch of disconnected worlds that don't communicate with one another, then the gravity of that metaverse will be less powerful than it otherwise might be if all participants were pulling together.

This is why I'm skeptical that plugging together dozens of existing games that were never designed to interoperate, all with closed loops of meaning, is likely to elicit much useful trade between worlds, even if we can overcome the technical problems. While it might be amusing to imagine Lord Voldemort fatally wounded by a hobbit wielding a machine gun from the *Halo* universe, such an event would likely break the predetermined systems of value in each of those universes. (It would also be a very strange day at Hogwarts.) You can get a better sense of how these metaversal interconnections might work by considering real-world culture. Fashion, sport, and music already interoperate, in a sense, so one could imagine virtual experiences rooted in these things being far more fertile ground on which to build networks of meaning. One could also envision games and intellectual property that are "native-born" to a connected metaverse being far more fruitful in supporting this transfer of value.

The ingenuity that powers a metaverse isn't like the creativity that powers the normal world of art and culture. You do not have infinite degrees of creative freedom in a metaverse, because every change you might make could potentially impact millions of other people. Ingenuity differs from creativity because it is rooted in the idea of solving a problem. Creating tangible value and useful shared experiences within virtual worlds requires the creator to work within the existing rules of the world. A creative work seeks to entertain or inspire or move a passive recipient. A work of ingenuity calls the recipients in and invites them to help advance and expand the parameters of the work. Ingenuity is creativity within moving boundaries, co-created by other creators.

The ingenuity that animates a metaverse requires its participants to create agreement, within this system of rules, that something new has happened or something valuable has been created. That act is a different kind of creativity than simply descriptive or representational creativity; it is one that gives human beings a very powerful sort of agency over society, over one another, and over the world. It's not that the metaverse grants you the power to *explain* the world, but that it grants you the power to *shape* the world.

In order to change the metaverse, you do it with problem-solving, not just pure creativity. We know this is true because it's been true across all metaverses throughout human history. At the dawn of the Roman Empire, after Julius Caesar was assassinated, the Senate and the public elevated the slain tyrant to godhood. In *The Twelve Caesars,* Suetonius wrote that the Senate issued a decree "in which they had bestowed upon [Caesar] all honors, divine and human," and that the common people of the Roman world thenceforth took every opportunity to advance the story of Caesar's godhood. "For during the first games which Augustus, his heir, consecrated to his memory, a comet blazed for seven days together, rising always about eleven o'clock; and it was supposed to be the soul of Caesar, now received into heaven," offered Suetonius as an example.

This historical example neatly illustrates the difference between creativity and ingenuity within a metaverse. To change the other world of heaven and introduce a new god was not something anyone could do. The maneuver involved the power of the Senate and the ultimate consent of the people. Someone who proposed such a change had to work within the existing rules, even perhaps creating evidence of miracles or portents.

This method of worldbuilding became a common exercise for the Romans, who elevated dozens of people to godhood over the span of the Western empire—justifying most of these deifications with ingenious readings of natural phenomena; constructing temples and statues that brought communities together to worship; and creating art that expanded the boundaries of these divine stories. Suetonius wrote that when Augustus Caesar was nearing the end of his life, for instance, a lightning bolt struck the letter "C" in "Caesar" on one of his statues; this occurrence was interpreted to mean that, upon Augustus's death, "he would be placed amongst the Gods, as Aesar, which is the remaining part of the word Caesar, signifies, in the Tuscan language, a God." Though the emperors were the ones being granted divine status, their continued divinity ultimately rested with the countless individuals who chose to put and keep them in the heavens, and who used the stories of these deified mortals as an opportunity to create meaning and value on Earth. When the Roman Empire fell, then so too did the deceased emperors' godhood.

If the ancient metaverses that we've been discussing had been fully, linearly real—in the sense that the planets Mars and Venus are real—then in some ways they would have been much less useful to the people who built them. If Zeus and the other Olympian gods, for example, remain exclusively far-off and distant figures who cannot be directly interacted with or seen, then their distance leaves a lot more room for their adherents to devise their own ways to praise, worship, and apprehend them—often through stories, songs, works of art, organizing principles for human society, and other meaningful and long-lasting cultural products.

But what if Zeus were visibly real? What if he popped down to Athens every Tuesday and demanded to be worshipped? I would argue that his visibility would actually hamper the real utility of having a mythology. The ability to interpret portents and create social change with reference to the gods would no longer exist for that community. One might imagine, somewhat cheekily, that even with real gods, people might well invent *other* gods to serve this important social purpose.

Likewise, if the metaverse were just a video game or one or more circumscribed worlds, then there would be no need for participants to create experiences therein or to expand its parameters. The developers would just build everything, and this top-down control would produce a more seamless user experience. But, in that scenario, the developers would also end up dictating the arc of meaning and realizing the bulk of the value. There is nothing particularly metaversal about that.

More than just a single legend, or tradition, or experience, a metaverse is an ingenuity matrix that binds together all of the constituent parts of its conversant worlds, keeps them stable within their own histories, and facilitates the exchange of value and meaning between the respective components. This bridge, this exchange, the breadth, depth, and strength of the connection: This is the conduit for value within a metaverse. As this connection grows stronger and more resilient, it conducts more meaning and consequence, facilitates more experiences, and empowers more people to expand the parameters of the connected worlds.

These, finally, are premises that we can work with, ones that we can use to assess the value of a metaverse from various standpoints, while avoiding false starts and negative externalities. A maximally valuable metaverse will allow for the free creation and flow of value within and between worlds, wherein the value derived from an experience in one world will also be consequential in the other worlds. It will facilitate ideation and ingenuity from a broad range of participants, rather than consolidating the experience-creation and meaning-assignation process in the hands of a chosen few. It will be conducted on democratic principles and invite democratic engagement, which will end up being better both for the metaverse itself and for society at large. The more people who can create and retain value within the metaverse, the more people there will be who want to use it. That, to me, sounds like a metaverse worth working toward.

DEFINING THE METAVERSE IN TWO PARAGRAPHS

Let's now collect these reflections into a fully formed definition of the metaverse. A metaverse is a collection of realities, including the real world or a "home reality" and a series of other worlds that a society imbues with meaning. Events, objects, and identities can exist in and be modified by multiple worlds in the metaverse. The utility of a metaverse lies in its ability to facilitate meaningful, fulfilling experiences in its constituent worlds. Value is transferred between worlds in many ways, including through increased social cohesion, the creation of valuable artifacts of culture, and direct commerce. A metaverse need not involve technology or physically embodied other realities accessed through VR or immersion and the like—but these additional ways of manifesting other worlds can enhance their value if they create more powerful ways for people to have fulfilling experiences, or to participate in the commerce in ideas that sits at the heart of the metaverse. The interplay between worlds, and the associated ongoing creation and transfer of value, is the foundation for virtual society.

But there is one further crucial component. Operating from this definition, you could make an absolutely beautiful, photo-realistic, hyper-immersive universe that could also be entirely useless. If this other world doesn't enable a participatory network of meaning that generates events or objects or experiences that are worth transferring between worlds, or engage a high proportion of society, then it will be functionally worthless. Conversely, a world which, for the sake of argument, was as graphically basic as *Habitat* but offered a strong foundation for a richly simulated world that allowed individuals and communities to seek fulfillment and generate value would be of enormous financial worth. As I noted in Chapter 4, *Minecraft* features very basic graphics, but nevertheless has become fundamental to the social lives of an entire generation of children.

A COMPARATIVE SCALE FOR METAVERSES

Now that we've defined a metaverse—and defined how value is created within a metaverse—we are faced with a very exciting opportunity: the chance to chart a course for a society that wholly embraces the value of a metaverse. How do we qualify the stages of maturation of such a virtual society, and how might we reasonably predict what the future could hold? I'm a big fan of the Kardashev scale of civilizations, devised by Soviet astronomer Nikolai Kardashev. The Kardashev scale tries to dramatically simplify the process of evaluating the advancement of a society by looking at its ability to harness energy. Kardashev civilizations come in three types, beginning with present-day society and culminating with a theoretical civilization that's able to harness the entire energy output of all the stars in a galaxy.

Inspired by Kardashev, I'd like to propose a different scale, one intended to simplify how we might understand the progression of a civilization's ability to create other realities that can improve people's lives. While we can't know specifically how a civilization might implement its version of a metaverse, we can predict the big milestones of its development trajectory. Let's define the levels of the scale as follows:

A *Level 1 virtual society* is what we've experienced until now: a civilization in which other realities are explored purely in words and ideas. Rich universes of religion, sport, and culture offer avenues for enhancing the meaning of individual lives while creating entirely new dimensions of the economy and improving social cohesion. In these societies, ordinary people can add to the narratives of these other worlds, but meaningful participation is mostly restricted to the elites. While you can't have a viable religion without thousands of acolytes, ultimately it's up to the priests to conduct the prayer services.

The dawn of mass media allowed us to visualize these other worlds and tell immersive stories about them. Interactive forms of entertainment can make it feel like these other worlds actually exist and actually matter, but the consequences of anything that happens on the silver screen or in *World of Warcraft* are unlikely to directly influence your life. This society can "dream" of other worlds very effectively, but those worlds remain escapes because there is no substantive network of meaning linking those worlds and our own. *World of Warcraft* is fun, but it isn't part of society in any meaningful sense.

A *Level 2 virtual society* is where we can see ourselves going over the next few decades of development. Technical advancement toward the metaverse means that massive virtual worlds can exist not only as self-contained diversions, but as part of an interoperable economy of digital assets, identities, and experiences that can directly impact your real life. This shift will create profound changes in the structure of society. Everyday individuals will have more opportunities than ever to add to the stories of these other realms, make or lose real fortunes, and conduct important relationships—all within other worlds mediated by technology. In a Level 2 society, the real world is still of primary importance, but its social, economic, and political order are profoundly influenced by what happens "over there."

The members of a Level 2 virtual society would consider the lives of those at Level 1 to be dramatically poorer than their own. Over the course of this period of history, one can imagine better and better immersion technologies and ever more complex worlds—but the crucial boundary is that these are always purely digital realities, accessible only through screens and devices. A Level 2 society is still subject to the basic physical boundaries and needs with which we are familiar today.

The key factors that define a Level 2 virtual society are, in my view, economic ones. We'll get to Level 2 when a decent proportion of people are engaged in jobs that exist entirely in the context of the metaverse, and when goods and services comprising things that exist mostly in that other world become a major part of the economy. This is the key difference between this stage and Level 1. In Level 1, a society might have many complex virtual worlds, but those worlds are not seen as central to the economic life of the real world.

A *Level 3 virtual society* occurs when a large number of people can tangibly travel to a simulated or constructed reality and live there—literally and fully. This physical inhabitation could be achieved through brain-computer interfaces, by having actually been born as computer code, or by some as-yet-unimagined exotic means. The *how* does not matter. What's important is that this method of existence will wholly transform the society that embraces it.

This civilization is one that lives continually in a metaverse of its own making—and what a life it will be! Time itself might run at different rates in different worlds, for example. Some realities might process hundreds of years in the time it takes others to experience a single year. You might actually live in the cave of the dragon you slew years ago, in a world where that occurrence matters deeply to millions of people. This age will be one of enormous possibilities. A Level 3 society's output of culture, its economy, even its sheer number of individual inhabitants is unimaginably vast compared to that of a Level 2 society. I'll also note that a purely digital existence would require much less energy than a physical one. In a Level 3 society, a world population of trillions of people would not be out of the question.

A single individual in a Level 3 society could live thousands of parallel lives. Groups of individuals could coordinate or create in ways that would no longer be restricted by physical limits. This society would be one with billions of times the cultural output, experiences, and individual opportunities as all the resources of the whole solar system could ever conceivably sustain in the real world, even if we drank the sun dry of energy and stripped every rock of matter to make new physical goods and locations. This society would likely perceive the slow, limited real world as a place to gather resources to power its vast corps of other possibilities. Crucially, the opportunities for individual fulfillment here are limited only by the availability of energy to power new experiences.

This vision, while eventually achievable, is pretty far away from being something we can build on Earth, and as fun as it might be to speculate about what it would be like to live in a Narnia-style metaverse—where you can go to the other world through the wardrobe, be entirely present there, and then return to Earth in time for supper—it's not what this book is about. While I'll return to the vast implications of Level 3 in the final chapter of this book, the Level 2 virtual society is what we're going to build first, and in the next three chapters I'll go into exactly how we might get there. In Chapter 6, I'll talk about how we might build the metaverse, and about the ways in which the stakeholders in a metaverse might grow, nurture, and encourage the network and its participants toward the goal of generating as much meaning and value as possible for everyone involved.

Chapter 6

BUILDING A VALUABLE METAVERSE: THE EXCHANGE

Though I might be subverting the premise of both this chapter and this book by saying so, I have to acknowledge that it isn't actually all that hard to set a virtual world in conversation with our own. We've been building forms of such worlds for eons, after all, and even in our current era we have found ways to make things happen within digital structures and then feel the ramifications here on Earth; the outcomes of all of the real-world elections affected by online disinformation are proof enough of that contention. But the challenge we face as a society isn't just to build a metaverse through which literally anything can flow, like a polluted river that carries equal measures of fish and garbage. The challenge, instead, is how to build a broadly valuable metaverse, one that improves the worlds that it serves, much like Heracles rerouting the Alpheus and Peneus rivers to clean out the fetid Augean stables. This task, too, will be a labor fit for Heracles.

In the last chapter, I defined a metaverse as a network of meaning and consequence that connects our world and one or more other worlds of living ideas, and allows for the creation and transfer of value within and between them. The development of a digital metaverse will be catalyzed by a combination of creative and market forces that will drive a network of creators and investors. The members of this network will come from a wide variety of backgrounds, and each will bear a special responsibility for the various tasks necessary to build a valuable metaverse.

The foundational task for the creation of a metaverse involves infrastructure: the software and hardware that will make virtual worlds run and will connect them with one another and the world. This task will cost money—lots of money. Individual and institutional investors will inevitably fund the early technological development of the metaverse and its infrastructure. Startups, established tech companies, and individual developers will do the work to create and refine that infrastructure. They will all seek returns on their investments, and there's nothing wrong with that; the trick will be to make sure those investments aren't repaid with disproportionate and stifling control over the metaverse.

Next, a valuable metaverse will need content: the constituent parts of the experiences that will define the best virtual worlds. Existing intellectual property holders will find ways to incorporate their creations into these emerging worlds, while artists, writers, musicians, filmmakers, and innumerable other creatives will find the metaverse to be an expansive palette for new work. Individual users, meanwhile, will inevitably advance and expand the parameters of the metaverse, and the experiences available within, in ways that we can neither predict nor control.

Just as important as content will be services. Small businesses and entrepreneurs will make the worlds of a metaverse more useful and functional by offering goods and services that ease and enhance users' experience of these worlds. Over time, as a metaverse grows in size and importance to its users, these businesses will come to constitute an economy that's as real as the real-world economy. Finally, the variations in content and services will create and inform the all-important differences between the worlds in a metaversal set. A metaverse would be a drab and dull place if every world within it looked and felt exactly the same. Different worlds will invariably have different visual identities, social customs, group priorities, and points of view. The sensibilities of these variegated worlds will be created by their users, and will emerge organically if those users are allowed and encouraged to make these worlds their own.

Any given metaverse is and will be the product of the choices made by the people who join forces to put it together. Some of these choices will be rational ones, and some will be irrational. Some will be made by independent actors, and some will be made under corporate direction, or under the aegis of some grand unified effort. While we can hope that most of these people will proceed with the goal of a broadly valuable metaverse in mind, some of the choices made will likely be malicious and selfish ones, premised on bad definitions, or made by people who are primarily interested in building a metaverse that they or their companies can exploit for profit.

This chapter is for people who are interested in building and experiencing the sort of metaverse that can maximize its social, economic, and psychological value. I'll offer some thoughts and ideas for how to increase the chances that the choices these people make are good ones. First, I'll discuss how the progenitors of a valuable metaverse will be more like gardeners than construction engineers, nurturing and tending a growing ecosystem rather than brute-forcing a structure into existence. Next, I'll discuss various options for how best to coordinate the network's emergence. After that, I'll describe the component parts of a valuable metaverse, talk about the people and entities that would likely be responsible for each part, and offer a vision for how they might start to fit together.

A metaverse, as I've mentioned, is akin to a game that a society plays together, albeit one that has real consequences and is designed for fulfillment, not escapism. It's a productive game, built around the concept of a constructed world of meaning. The basis for this productive social game is that we all agree to treat it as if it is real. The content and fulfillment found in a metaverse will be delivered by people acting out of individual self-interest, coming together to build and expand worlds by adding their own components to them, one by one. A metaverse will change people's lives for the better if and only if it allows all participants the chance to hold a meaningful stake in its health, growth, and success.

EMERGENT COMPLEXITY

In practical terms, a metaverse is going to be a set of massive, complex, real-time simulations, which in turn are mediated by an economic and social layer, which itself will incorporate mechanisms for the storage and transfer of value, such as nonfungible tokens and blockchain. The scope of work required to build a metaverse like this, let alone a truly valuable one, will be vast and interdisciplinary. We'll need great programmers, designers, and engineers, yes, but we'll also need experts in economics, organizational behavior, social behavior, and ethics. We'll need established artists and emerging artists. We'll need to obtain the cooperation of governments and elected leaders. And once these stakeholders are identified and onboarded, we'll still need to build the damn thing.

Generally, when you set out to build something functional, it is best to work from exact specifications. Whether you're building a house, a car, a computer, or a piece of Ikea furniture, there isn't all that much room for improvisation in the building process if you want these things to work the way they are intended to. Instead, you use a blueprint, and you follow that blueprint to the letter. (One need only google "Ikea fails" to see how quickly and hilariously things can go wrong when you don't.)

When it comes to the infrastructure of the metaverse, we'll have to deploy specific guidelines. It's unwise to be imprecise or vague when it comes to technical specifications, so there must be very clear protocols and communications regarding the hardware and software that will power the metaverse. (Exactly who might be setting these protocols and sending these communications is a question I'll explore later in the chapter.) These are the areas where you don't really want to "wing it," because winging it creates serious problems in terms of usability and interoperability.

The infrastructure for the metaverse will have to be monolithic, by which I mean that there won't be much room for community input on how to lay the cables, so to speak. Someone will have to build it, and someone will have to pay for it—and, as I mentioned earlier, these people will likely want some return on their investments. Beyond the infrastructural level, though, the metaverse won't be built so much as it will emerge, like an artistic movement, in ways that cannot really be predicted or controlled. The goal for any group seeking to create an optimally valuable metaverse should be to organize it so that those who foot the bill for its infrastructure do not also feel entitled to dictate the content of the metaverse, or to hoard all of the value that emerges from it.

If the pathway to infrastructure is narrow, the precise opposite is true when it comes to the metaverse's content. The programmer John Carmack has observed that if you try to make the metaverse, you will fail. What I think he means, in part, is that an elaborate blueprint would be both unnecessary and misguided when it comes to creating the conditions for the culture and content of the metaverse. Instead, a digital metaverse will emerge iteratively and unpredictably, out of a general cultural ferment and conditions conducive to its growth. A valuable metaverse will take shape organically as a product of the choices made by each individual actor therein.

The philosophical concept of *emergent complexity* broadly posits that the elements of complex systems—ones in which the constituent parts interact with one another and the environment in ways that are hard to definitively model—will organically and inevitably organize themselves in patterns, groupings, and interactions that could not have been predicted at the system's outset. In 2015, for instance, players of the open-ended simulation *Dwarf Fortress* started to find a surprising number of dead cats within the world of the game, often covered in their own vomit. It took a while for the players to realize that these cats had been walking across the floors of virtual taverns on which beer had been spilled; when the cats later licked their paws to clean themselves, they ended up ingesting the beer, getting drunk, and dying of alcohol poisoning. A small number of actors in a system, each behaving autonomously—occasionally coordinating their efforts, generally being influenced by one another, and growing and changing as a result—tend to produce unpredictable patterns and outcomes that are revealed only upon taking a macroscopic view.

When you look down at a city from the windows of an airplane, the rhythms and relationships of the streets, buildings, and neighborhoods reveal themselves in remarkable ways. When you view your garden from the street, you perceive things that you cannot see when you are focusing on pruning one specific tomato plant. As the individual components of a system increase, additional complexity emerges. Systems grow deeper and more dense with meaning when their components are free to interact autonomously with one another, and to determine their own directions for growth.

Emergent complexity exists in tension with behaviorism, a philosophy that, in its strictest interpretation, believes that maintaining strict control over a system's inputs will lead to strict control over its outputs. It also runs counter to the production dictates of our current age, which often require certain tasks to be completed a certain way by a certain time, with little tolerance for improvisation within the process. But emergent complexity is nevertheless the soul of our most interesting software applications, and virtual society will not come about without it.

Software users often end up using sufficiently complex applications and programs in ways that the developers never intended or predicted. Take Twitter, for example, which began as a means by which to broadcast your physical location via SMS to small groups of friends, and grew into a worldwide microblogging service that, for better or for worse, has become integral to the process of politics and governance. Or take the gaming practice of "speedrunning"—the act of zooming through digital game levels as quickly as possible—around which a robust community has grown, with speedrunners competing against one another to see who can log the fastest time. The people who programmed *Super Mario Bros.* back in the 1980s likely did not expect that, decades later, a subculture of players would take joy in racing through the game's levels like Olympic sprinters, yet here we are in a world where the speediest players can beat the entire original Nintendo game in a mere 4 minutes and 55 seconds. It's fair to assume that if a software product is sufficiently engaging, its users will disregard the developer's ideas of how they ought to use it, and will instead come to use it in ways that serve their own interests and needs.

An optimally valuable metaverse will emerge in a similar fashion. It will not be a finite product to be controlled and dictated in a top-down manner, as if the developers of a metaverse are its auteurs and the rest of us merely captive to their vision. Unlike a fixed story bounded within the covers of a book, a metaverse is a set of ideas in which the ideas themselves have agency. After laying down the initial rules, the "author" of a metaverse doesn't decide what happens next: We all do.

This does not mean that there's nothing for the organizers of a metaverse to do other than lay the cables and wait for the worlds therein to populate themselves with meaning. To extend the gardening metaphor a bit further, there are differences between a weedy vacant lot, a topiary sculpture, and a bountiful garden. A vacant lot becomes overgrown and disused because nobody tends it, and the ensuing tangle can make the lot a blight. A topiary sculpture, on the other hand, is an example of one or more people forcing nature into a shape that it would not have taken on its own. While some topiary shapes are more appealing than others, all of them are the product of an individual auteurist vision.

A bountiful garden, though, grows according to its own logic in a manner that can be both beautiful and productive for lots of people, not just vagrants and autocrats. It is tended, not controlled. If the organizers of a metaverse want to maximize its potential value, they will have to find ways to tend it so that value emerges, rather than shape it so that value is forced in a certain direction or abandon it so that no value emerges at all.

THE CASE AGAINST THE CORPORATEVERSE

Like the pyramids, or Göbekli Tepe, a broadly valuable metaverse has the potential to be a civilization-defining project. But the pyramids took centuries to build. Göbekli Tepe took a thousand years. Building a valuable metaverse will also be a massive, challenging task, and likely an intergenerational one, too. Who will coordinate all of this work? Who will help the various parties work together and the various components fit together? And how do we create and sustain societal buy-in over a long period of time?

When considering organizational structures for the development of a metaverse, we must ask ourselves the following question: Which structure is most likely to facilitate the creation of the most value and meaning for the most people over the longest term? At any given point in the process, the developers of a metaverse will have to prioritize within technical and social constraints, which will inevitably mean tradeoffs between immersion, presence, accessibility, fidelity, and the bandwidth of meaning within and between worlds. How do we ensure that the tradeoffs we make add or maintain value rather than subtracting it? And who are the "we" that I'm talking about?

Already, certain big corporations have trumpeted their visions for the metaverse, the implication being that those companies will assume the leadership role in building it and the ownership role in controlling it. This model—let's call it the Facebook model, though you can replace "Facebook" with the names of most large tech companies and get similar results—is perhaps the model of the metaverse with which you are currently most familiar. Facebook has been very vocal about its plan to build and dominate the metaverse in a similar manner to how it wields great influence over the internet. By getting out in front of the discourse, the company hopes to inextricably associate itself with the metaverse, and in turn make its centrality to and oversight of it a foregone conclusion.

This organizational structure for the development of the metaverse would result in a metaverse that resembled a topiary sculpture, in which a relatively small number of people would dictate its shape and meaning, while the rest of us would be limited in the extent of the value and meaning we could add to or take from it. The economic value of this sort of metaverse would disproportionately consolidate around the platform providers, leaving a meager share to trickle down to everyone else.

The notion of one big company spearheading and subsequently controlling the metaverse feels familiar, because this is how the internet works today. The modern internet is dominated by a handful of massive, vertically integrated companies that own all of the data generated by the users of their platforms. User data is the product that these companies sell, in many cases—hence the aphorism that, for Facebook and Google and many other similarly organized companies, the user is the product. The more user data a company stores and controls, the more valuable that data becomes, which is why it is so important for these companies to keep growing their platforms—and why they make it so hard for their users to leave those platforms, or to exercise any meaningful autonomy within them whatsoever.

Individuals aren't the only users who struggle to exercise autonomy on these platforms—businesses, too, are disempowered. The platforms control such a significant percentage of the revenue generated therein that, even if it were technically possible to do so, there really would be no incentive for any founder or startup to spend the amount of money necessary to build many types of businesses on those platforms. Individuals, of course, run small businesses on these platforms all the time, and there isn't much initial investment required if all you want to do is, say, resell sneakers on Facebook Marketplace or produce videos to post on YouTube. But the initial costs of founding an ambitious tech startup are such that those costs will never be proportionally recouped if the business is tethered to and limited by a larger platform. You can build businesses on the infrastructure, and you can use social platforms for marketing, but most big opportunities are stifled, or just turned into features and copied by the biggest companies.

The "own the platform, own the user" model has become the one that every other tech company feels it must follow if it, too, wishes to one day become a Silicon Valley megacorporation. The corporate psychological legacy of Web 2.0 is that a staggering number of founders and investors are convinced that monopoly is the lone path toward prosperity. But this divisive, controlling model won't work as well with the metaverse because of the very nature of what a metaverse is. The corporate model of the metaverse would almost certainly not be one in which the community of users would feel any stake in or responsibility for the health of the ecosystem—and it definitely wouldn't maximize the metaverse's potential value.

The value structure of most of the platforms on the internet today is shaped like a pyramid. The platforms make up the base of the pyramid, which is where the bulk of the value resides. The creators are up at the tip of the pyramid, and the value that they take is proportionately smaller. Creators rely on these platforms and their network effects to such an extent that most of them are effectively held captive by these platforms. Even the highest-earning YouTube creator or Instagram influencer is still small potatoes, relatively speaking, both because the amount that these people earn is minuscule when compared with the amount that the platform is making, and because very, very few individual creators or influencers would ever even be able to dream about starting their own competing platforms. The social influencers Jake and Logan Paul aren't going to leave YouTube for PaulTube, unless PaulTube is some pay-per-view boxing thing distributed over existing networks. Despite their massive popularity, they just wouldn't have the resources necessary to do it. Unless you somehow possess truly generational wealth, chutzpah, and business talent, going fully independent of the big social platforms is an Icarus maneuver for creators.

But the metaverse will work on a different basis. If you accept that the metaverse will be composed of embodied three-dimensional worlds, and that most of the experiences found therein will both look and function similarly to the way a good computer game looks and functions, then you also must accept that the act of creating these experiences will be much more resource-intensive than, say, posting a funny meme to Instagram. Yes, individuals will be able to create and provide lower-stakes, lower-fidelity experiences that won't require much capital investment. But the users of a world will also hunger for more elaborate experiences. Even if the level of work required to make an elaborate experience in the metaverse was equivalent to the level of work required to make a good video game, a developer would still rack up tens of millions of dollars in costs, at least.

YouTube can largely leave the content production to its creators because making videos is cheap. It might cost as little as forty-seven cents to make a viral video, which means that the barrier to entry is so low that pretty much *anyone* might envision themselves making a viral video. But because of the amounts of money involved in experience creation within a metaverse, platforms will have to incentivize companies to build businesses on their platforms. In that case, the platform can't be shaped like a pyramid, because those companies will see that they won't have much chance to recoup or profit from their investment.

In order for a metaverse to populate with the quantity and quality of worlds and experiences necessary for it to be worth anyone's time, then, it will have to resemble an *inverted* pyramid, where the infrastructure providers take the smallest percentage of value, and the rest of the value is created by and accrues to the creators. Otherwise, the creator of the metaverse will have to foot the bill for all the content—and even Facebook can't afford to spend $200 billion per year commissioning exciting metaversal experiences.

There are certain merits to the corporate model. From a usability standpoint, a corporate-built metaverse would be very likely to just work, for one thing. Usability was one of the social Web's key selling points. The corporateverse wouldn't be very buggy, and there would be a certain seamlessness to the user experience. Big companies could and would leverage their existing infrastructure to make the user-side transition from the internet to the metaverse an intuitive one. Because these companies are already very well capitalized, they wouldn't have to struggle to raise the money to build the metaverse from scratch.

They could just *do* it, and they could do it more quickly than most other entities could.

But even if companies such as Facebook could build a metaverse quickly, I do not believe that they would do it well. At the time of writing, Facebook's understanding of the bandwidth of meaning between the various worlds centers around convergence—i.e., the "you" in the metaverse is the same you as in the real world. You'll likely use your real identity in the Facebook metaverse, which means that the company's vision for the metaverse is fundamentally devoid of anti-structure. Their vision for the metaverse also emphasizes immersion rather than presence, and is fixated on building VR headsets. And yet, as I noted in Chapter 4, developing and populating an immersive virtual environment that is broadly indistinguishable from real life has very little to do with true value within a metaverse.

People want fulfillment from their virtual experiences first and foremost, and in a truly valuable metaverse these fulfilling experiences can and will be mediated by all sorts of interfaces and all sorts of devices. Will you eventually be able to connect to the digital metaverse using fancy futuristic goggles? Probably, yes, but you won't need to. We can have an amazing metaverse that does everything I've talked about in this book without ever creating fully immersive experiences featuring VR headsets that will cost you a lot of money to acquire.

It is instructive to contrast Facebook's vision for the metaverse with the vision set forth thus far by Epic Games, creator of *Fortnite*. In Epic's vision of the metaverse, the company is targeting presence, not immersion. *Fortnite* is very focused on accessibility, which is good. But it, too, is presiding over an unbalanced network of meaning. While *Fortnite* is bringing brands into its world, nothing of value comes out. The events and experiences that happen in *Fortnite* don't matter in the real world. In April 2020, the rapper Travis Scott held a highly touted concert within *Fortnite* that got a lot of publicity—but, ultimately, the concert was a novelty. Yes, most experiences in virtual worlds will begin as novelties before they're ingrained as value systems. But novelties generally don't become meaningful in the absence of a clear and easy conduit for value transfer. Nothing that happens in *Fortnite* actually matters in the real world, because the world of *Fortnite* is not fundamentally organized to interface with the real world, or to return value and meaning to the real world and other businesses.

Because of the superficial similarities between great games and virtual worlds, it's easy to presume that a great game might easily evolve into a full-fledged virtual world, and from there expand into a metaverse. This path will be harder to follow than one might think. For one thing, there will be serious technological challenges along the way. Technology built for one specific purpose tends to be inelastic in its ability to support post-hoc purposes. A car and a speedboat both have engines that consume fuel and help you get around, but you'll quickly realize the difference between the two as soon as you drive your car into a lake. A game developer that just wants its game to become a metaverse will be quickly beset by tech flaws and setbacks, because a game and a metaverse present two different imperatives with two different sets of needs and protocols. Instead, the game developer will have to build a bigger structure—and they'll have to build it from scratch.

In the cases of both Facebook and Epic, these metaverses are optimized not for fulfillment and utility, but for profit to be taken by the creators of the platform. The company that built these versions of the metaverse would likely also be the entity that created and managed the bulk of the experiences found therein. Without the ability to support and sustain a panoply of individual creators, without the will to build these virtual worlds around principles of fulfillment and individual utility, a corporate metaverse would be a metaverse in name only. Sure, it might be visually immersive, but it would nevertheless be psychologically and spiritually barren.

Because the big companies that built these metaverses would also own and control them, any stabs toward democratic governance therein would feel grafted on, superfluous to the primary goal of enriching the company's officers, investors, and shareholders. In the big-company model, individual creators would necessarily be disempowered, because they would hold no real equity in the experiences they're creating, and no stake in the broader health of the ecosystem in which they'd be creating them. Individual users, and the data they create, would also be at the mercy of the people who control that data and who will inevitably package and sell it to advertisers. But no matter what companies may want you to think, a corporate-controlled metaverse isn't a foregone conclusion. There are other ways.

THE ANARCHY MODEL AND ITS DISCONTENTS

The polar opposite model for bootstrapping a metaverse is what I call the "anarchy" model. In this model, hackers, individuals, startups, nonprofit groups, and anyone else who wants to do so will throw cohesion to the wind and end up building the metaverse themselves, with no central entity tightly controlling their labor or the project's development. Picture the early days of the World Wide Web, where the network was populated by innumerable idiosyncratic personal websites, none of which really fit together in any meaningful centralized sense. Or picture the open-source software movement, in which idealistic individuals worked together to create operating systems and other programs, the development of which was motivated not by dreams of personal profit but by shared belief in an orienting philosophy. These products and projects stood in opposition to the notion that one should seek to extract financial value from internet or software applications.

This organizational model is the most idealistic one, and, if it came to fruition, it could produce a metaverse with a very wide range of interesting experiences. There is already some activity on this front. Decentraland, launched in 2017, bills itself as the first-ever virtual world owned by its users, in which digital land is commodified as NFTs and purchased by means of cryptocurrency. Writing in *PC Gamer* in March 2020, Luke Winkie described Decentraland as "*Second Life* meets libertarianism," a world in which "every piece of content in the game is owned, completely autonomously, by the players."

The Decentraland model—speaking generally, not specifically—has its heart in the right place. But, for lots of reasons, this model is as likely to produce a failed product as is the corporate model. If, per our gardening metaphor, the corporate model is sort of like a topiary sculpture, then the anarchy model would be more like a vacant lot, where growth is both unimpeded and unplanned, and thus ends up inhibiting use and the creation of value and meaning. A corporate metaverse would suffer due to excessive top-down control and coordination. But the metaverse produced by the anarchy model would suffer because there would hardly be any coordination at all. While too much corporate money might spoil the metaverse by making it too commercial and profit-focused, if you're trying to build the metaverse strictly out of a can-do spirit of volunteerism or a sense that no one should tell you what to do, then guess what:

You're never actually going to build a functional metaverse. The end product will not be directly useful for anyone other than the hobbyists and specialists who built it.

We've already seen this sort of noble dysfunction with Decentraland. Though the platform is an interesting experiment, as of this writing it doesn't work particularly well. "It is a rickety product," wrote Winkie, noting the world's "frame rate hitches," "weird screen-scaling bugs," and "brutally long loading times." Worse than the technical glitches is the fundamental emptiness of the world. "I truly didn't see a single other soul during my time in Decentraland," noted Winkie. "Nobody, from the museums to the pirate's cove."

The anarchy model risks producing a bunch of worlds that no one actually uses; a collection of walled gardens that do not interoperate, between which meaning and value cannot be easily transferred. The ensuing metaverse would be like a world filled with cloistered nations that could be entered or exited only with extreme effort, where the currency from your home nation could neither be spent nor exchanged in any of the others. While we don't want the constituent parts of the metaverse to be polished to a lifeless sheen, we also don't want them to not fit together at all. For a metaverse to work to its full potential, it must manifest the sort of seamlessness that is difficult to ensure when you're working primarily with volunteers and individuals whose labor isn't compelled by salary or coordinated by some broader entity.

So, if anarchy doesn't work, and top-down corporate control doesn't work, then what organizational model would work? It would have to be a middle ground, one involving cooperation between many entities and input from both corporations and individuals. The middle-ground model creates space for a wide variety of inputs and perspectives, while leaving room for the sort of project-management roles necessary to make the metaverse happen. I believe that this is the optimal organizational model if you hope to create a truly valuable metaverse. I call it the Exchange.

THE EXCHANGE

An exchange can be a place, an activity, or a philosophy. It's where you go to buy and sell things, to trade value; it's where you go to share ideas and conversation and experiences. Implicit in the very notion of *exchange* is that value is bilateral. You don't go to an exchange to hold on to something, you go there to put it into the world. An exchange is a bridge, across which meaning and value and consequence can flow and develop and transform and evolve. This image is a relevant one around which to construct an organizational model for the metaverse.

In this model, an array of stakeholders, likely dozens of them to start, would come together and create a consortium, involving representatives with expertise in technology, business, game design, ethics, politics, media, the arts, psychology, and so on. Admission to the consortium would be offered first by invitation—the people doing the inviting would probably be the small handful of "founders" who came up with the Exchange in the first place—and later by application. Membership would be contingent on agreement with the central goals, principles, and definitions of a valuable metaverse. Agreeing to abide by this code of ethics would differentiate this group—which you might think of as a board of trustees, or a professional society— from any other small group of people hoping to build the metaverse. They would come together not for reasons of boundless personal profit, but for altruistic purpose. The members of this group would pool their resources and expertise to create the financial, technical, organizational, and ethical conditions necessary to assemble the constituent parts of the metaverse. (I'll note that this point isn't purely theoretical; it's the model we've begun to put in place with the M^2 project, and early results are promising.)

What are the component parts of a valuable metaverse? As I noted earlier in this chapter, on a purely practical level, we'll need significant levels of technological and infrastructural investment. The operations-per-second requirements of a minimum viable metaverse will require a staggering amount of computing power and storage, which will in turn necessitate the globe-spanning infrastructure that can support and sustain the sort of computation required for metaversal applications. We'll need the right hardware—the storage, the power sources, the network infrastructure—and also a well-designed system built to meet its eventual users' social, economic, technological, and pragmatic needs. The Exchange could fund the development and implementation of these technologies while devising and promoting a set of standards on which they would run.

These technologies will power the development of virtual worlds and useful experiences within those worlds. We've talked a lot about worlds and experiences already, but they sit at the core of the metaverse and its psychological utility to the individual. These experiences must be geared toward intrinsic fulfillment, and the worlds in which they exist must be configured so that a wide range of people are empowered to create them and to derive value from them. They can and will encompass everything from fantastical heroic adventures to learning a pragmatic skill, from attending a virtual concert to sitting at a virtual bar and chatting with a bartender.

The Exchange could coordinate the development of these

worlds and experiences, assembling and resourcing the programmers, designers, and artists who will create them. But it
will be necessary for other people to be able to create and add
value within these worlds, too, and we'll have to incentivize
them to do it. A valuable metaverse will be one that makes it
easy for an individual to create experiences within a given virtual world. As such, the worlds in the metaverse must give individuals access to the data and tools necessary to offer these
experiences, and the ability to realize economic gain from their
labor. A consortium like the Exchange, which understands the
value that individuals will add to the metaverse, will be able to
ensure that the metaverse is built in such a way that allows them
to do so.

This leads to another component part of a valuable metaverse: We'll need a meta layer of social and economic value that
ties these worlds and experiences together. If there are clear ways
to create, store, quantify, and exchange this value, then these
worlds and experiences will carry meaning and consequences
that matter on more than just an individual psychological level.
This is where I foresee blockchain-style mechanisms coming
into play: intricate computing processes that serve as independent guarantors and clear ledgers of value within the world.

As I noted earlier, it can't be that all of the economic value of
the metaverse accrues to the developers, because that outcome
would perpetuate the inequity found in the internet today, as
well as stymie investment in and commitment to the metaverse;
people would feel no ownership over it. In both an anarchic and
a corporate metaverse, the creation of economic value would be
inhibited—in the corporate metaverse because the developing
company would reap most of the windfall, and in the anarchic
metaverse because there would be no clear way to create, store,
and transfer value between virtual worlds and into the real world.

If the Exchange were to oversee this process, though, it could ensure that the various virtual worlds comprising the metaverse will talk to one another—and that they will all speak the same language. It could ensure that the various worlds are built on a foundation of blockchain technologies that make it easy for economic value to be created, stored, and transferred. It could also help to create the vital and necessary bridge between the meta layer and the real world, which will require real-world structures to recognize the merit and meaning of the meta layer and work to integrate with it. There needs to be a way in which the value and meaning created in virtual worlds can affect the real world; there must be a lasting and meaningful connection between the various worlds. This is as much a social challenge as it is a technical one, and a well-resourced consortium would be ideally situated to advocate for these goals with real-world parties—banks, businesses, governments, service providers, NGOs—and negotiate with those parties to help create these conduits and keep them open.

Cryptocurrencies and blockchain-style technologies will be integral to any efforts to build and maintain a bridge for the transfer of value within a metaverse. They will be the guarantors of the individual self-interest that will serve to populate the metaverse with lasting meaning. We're going to have to rely on individuals and individual companies to create meaning within the metaverse, and in exchange for their work these parties are going to want to know that they can realize profit and recoup costs; they'll want to know that it is worth it for them to spend time and money developing and providing these experiences. For this to happen, a transparent financial instrument will have to be baked into the core of a metaverse, so that creators won't be at the mercy of the platform provider.

Ideally, the members of the Exchange will be able to interface with businesses and intellectual property (IP) holders to get them to understand the opportunities that will exist for them if they are able to adapt to this new world. The metaverse will be built by user-created content and user participation, and the most successful content creators will be those who are able to drop some of the deep resistance that storytellers and rights holders have around letting other people play with their IP. Take *Star Wars,* for example, and the ways in which Disney is reluctant to license the characters of that universe, in part because it wants to control those characters and retain possession of the stories told around them. Well, the metaverse will present different modes for creating and interacting with content. Everyone's going to be LARPing together in the metaverse, so to speak, and thus these worlds will have to find ways to devise new paradigms for creation and IP transfer that suit the needs of this new medium while still incentivizing creators. The challenge to the games industry is, essentially, to go from letting users *play* a game to offering them meaningful experiences that are not necessarily gamified, and to concurrently change the monetization strategy for games. The Exchange will be well situated to discuss and answer these questions in a broadly valuable fashion.

A valuable metaverse will also need a transparent method of governance and a means by which to encourage ethical, prosocial behavior within it. These two imperatives are linked. It's very important to devise and promote a shared set of values that will overlay the metaverse, and to build these values into the metaverse's core. If we can accomplish that, then the different players in a metaverse—governments and regulators, companies that may seek to build infrastructure or experiences, content creators and talented people, and so on—will be usefully constrained by these shared values.

Eventually, I believe that an optimally organized and governed metaverse is likely to end up a nation-state of sorts, or a new kind of state-like entity. I'll have more to say about this in Chapter 8. For now, though, it is worth noting that companies such as Google and Facebook have already assumed more power than many nation-states, and in some ways have started to act like autonomous countries. If these companies govern like autocracies, then a valuable metaverse must look more like a democracy. But democratic structures will only emerge from an Exchange-style organizational model for the metaverse. Whereas an anarchic state isn't a state at all, and a monopolistic state is a dictatorship, a state-like entity organized around the principles of exchange will both preserve and add value for the most people over the longest period of time. The Exchange, as I envision it, would be responsible for oversight and the promulgation of prosocial principles, but it would also have to avoid the urge to micromanage. It would have to be able to get out of the metaverse's way.

A CONTINUITY OF BELIEF

In order for that to happen, all parties to an optimally valuable metaverse must share a philosophical concordance: a shared investment in certain core values and in the core meaning of the metaverse project. This sense of the project's importance will have to be strong enough to outlive the consortium's initial founders and funders. I mentioned earlier that building a valuable metaverse would be an intergenerational undertaking. As we've seen with other intergenerational projects throughout history, there must be a continuity of belief that undergirds their development.

Take the ancient Egyptian pyramids, for instance: living

monuments to the power that a shared social belief in a meta-versal reality can exert on the real world. Each succeeding generation that worked on them, and/or existed in a world in which their construction was a priority, had to believe that the project was worth it—that it was not just logical but necessary to invest so much time, money, and effort into a longitudinal, impractical endeavor.

The question of how to create long-term buy-in for an ongoing sociocultural project that, on its face, might not appear particularly practical isn't exclusive to ancient Egypt. In 2011, the U.S. government agencies NASA and DARPA launched an initiative called the 100 Year Starship Project, with the stated goal of making interstellar travel a reality within a century's time. The main philosophical challenge of this project, just as with the pyramids, involves how to sustain motivation over the years. A lot will change over the course of the 100 Year Starship Project: Leaders will come and go, social priorities will evolve. The world we'll have at the end of the Starship Project will assuredly look very different from the world we had at its launch. So how do you get successive generations to sustain the necessary enthusiasm for and investment in the project?

The answer, in large part, lies in convincing people that the problem being posed is one inherently worth solving, even if it takes a hundred years to solve. In ancient Egypt, for example, forced labor and pharaonic decree could only do so much to get an entire society to agree that it made sense to spend centuries building a succession of expensive decorative triangles in the desert. A fantastical project of that magnitude had to resonate in some respect with the society itself, not just with its leaders.

Like the pyramids and the 100 Year Starship Project, the metaverse that I've just described is a project that will likely require billions of work-hours before it even comes close to full fruition—though it can create a lot of value along the way, and it can be built in stages. Creating a vibrant set of virtual worlds, and a virtual economy that links all of these worlds and the experiences available therein, will be a multigenerational endeavor. In order for society to justify the time and expense it will take to make the metaverse, we must all have a shared understanding of exactly why these other worlds will be worthwhile. We must proceed from shared premises, and those premises must be strong enough to create buy-in that will last through generations.

The Exchange can help promote these premises, and it can also help instill the ethical infrastructure necessary to protect against the prospect of the metaverse going sour and becoming an antisocial ecosystem. At some point, certain members of even the most idealistic consortium will look at the developing metaverse and wonder whether it isn't time for them to abandon the community and instead start looking out for their own financial interests. It's not so much a case of *whether* this will happen, but *when*. Before we get to that point, we must build a system that expects and can sustain individual selfish behavior without collapsing; a system that is scrupulously ethical even if every single individual node in that system is looking out for their own interests. We will need a system that is centered around value creation for everyone, not just the people who brought that system into existence. The best way to do this is to build a virtual economy that is bursting with fulfilling and lucrative virtual jobs.

Chapter 7

VIRTUAL JOBS AND THE FULFILLMENT ECONOMY

For a tuneful primer on one of the fundamental problems with the modern economy, you could do worse than listen to the classic Disney song "Whistle While You Work." The song suggests that the best way to escape the drudgery of daily labor is to find some means of distracting yourself from the fact that your tasks are unpleasant ones. Whistling takes your mind off what you're doing. Pretending a broom is your sweetheart helps you forget just how much you hate sweeping the floor. These sorts of coping strategies are necessary only in a world where the jobs we perform are inherently unfulfilling—where there is scant intrinsic satisfaction to be found in our work.

As I argued in Chapter 2, this is the world many of us inhabit today. Modern society is centered around principles of productivity. A central measure of a nation's economic prosperity is its

GDP, or gross domestic product; this metric literalizes the concept that the healthiest economies are those that are the most productive. From a young age, individuals are taught that it is noble to work long hours without complaint even at tasks that are dull or demeaning. We have been socially conditioned to believe that toil is a virtue, and the economy has taken advantage of our credulity. As production imperatives have intensified and the pace of automation has quickened, not only have we all ended up working more, but many of our jobs have also become much less satisfying.

From a narrow economic perspective, there is nothing wrong with this trend. Individuals' fulfillment levels don't directly factor into a nation's GDP, after all. While employers have few incentives to make their workplaces deliberately unpleasant, neither are they incentivized to organize them around humanistic principles. Indifference toward workers' intrinsic needs is often manifested at low-wage workplaces, such as Amazon's massive fulfillment centers, where employees are subject to regular quantitative performance analyses and are allowed little autonomy. Protests about these sorts of workplaces often emphasize working conditions and wages—but they rarely note the existential injustice in the fact that the jobs being performed are so often mindless. Instead, we have all just sort of internalized the notion that a job is supposed to be unfulfilling. As the saying goes, that's why they call it "work."

It's worth taking a moment to reflect on the strangeness of this standpoint. We only get one life, after all, and we spend most of it at work. Advanced technology hasn't loosened the stranglehold that our jobs have on our lives: It's tightened their grip. And yet we've somehow just come to accept that our jobs—which consume the bulk of our waking hours—are not at all obliged to meet our most basic psychological needs. Even those jobs that do optimally challenge their workers are often characterized by stress and overwork. As Yale's Daniel Markovits wrote in *The Meritocracy Trap,* many white-collar professionals have been trained their whole lives to make work *into* their whole lives—to accept that there is nothing wrong with being expected to bill ninety hours per week while never seeing their kids. Society tacitly tells us that we should not expect to find fulfillment at work, and yet the increasing amount of time we spend there can preclude us from seeking fulfillment elsewhere. There is a fundamental inhumanity to this construct.

Working a job can and should be a useful experience. When humanely organized and conceived, our jobs can give our lives a sense of purpose. They can provide us with optimal challenges and can confer a sense of pride, both in our own competencies and in participating in the economic life of society. And yet the structures and incentives of modern employment have instead elevated productivity into an end in itself. As I've noted elsewhere in these pages, this premise is unsustainable. On an ecological level, runaway production imperatives decoupled from the needs of the societies they are supposed to serve have decimated the environment via anthropogenic climate change. What better symbol is there of the malignance of the production ethos than its metastatic consequences for humanity's sole habitat? On a social level, this premise has created a widespread crisis of purpose that has come to destabilize the world.

In his book *Bullshit Jobs,* David Graeber explored one major reason why so many of us are unhappy at work: It's because our jobs aren't built with human fulfillment in mind. The only thing you get out of working a "bullshit job" is the often meager wage that you earn. A bullshit job is one that fails to engage or challenge the worker on an intellectual or emotional level; one that isolates the worker and deprives them of autonomy; one that is disconnected from any sense of broader social purpose.

The feelings of malaise and underfulfillment conferred by bullshit jobs are poisonous ones. We have seen how these feelings are especially keen among young, unmarried men—known as YUMs by those who study the social unrest for which this group is often the catalyst—and the underemployed middle-agers who, in the twentieth and twenty-first centuries, have flocked to reactionary political movements that promise to restore purpose and order to the world. While there are other, darker reasons why people support these movements, political upheaval often stems from individuals' disaffection with economic structures that seem to be rigged against them.

Over the last half century or so, productivity has skyrocketed while wages have stagnated, and people rightly perceive the injustice in this disparity. The philosopher John Rawls argued that, in a liberal society, justice is equivalent to fairness. In economic terms, fairness is often exclusively perceived as a matter of income inequality, in which wealth disparities create and perpetuate a system of *haves* and *have-nots*. But it is also fundamentally unfair and unjust that so many of our jobs fail to meet our most basic psychological needs. For the bulk of the industrial and post-industrial eras, employers have told us to whistle while we work: to use artificial means to distract ourselves from the fact that so many of our jobs are not built with human fulfillment in mind. Rich or poor, white-collar or blue-collar: When it comes to fulfillment at work, most of us are have-nots these days.

In this chapter, I will argue that the metaverse offers a promising solution to this era-defining crisis of capitalism. I'll explain why, as virtual worlds continue to grow in size, popularity, and meaning, they will inevitably end up creating a robust virtual economy rooted in principles of individual fulfillment. This economy, I'll argue, will produce millions of new employment opportunities—desirable, lucrative jobs premised on the production of meaningful experiences. I'll make the case that an optimal metaverse will lead us away from a production economy and toward a fulfillment economy.

The process of creating fulfillment in a virtual world is a symbiotic one. It's not just that the jobs in the metaverse can or will be more fulfilling than the jobs in the real world—it's that human effort is required to make the metaverse into a fulfilling space. Virtual worlds become real and impactful when a critical mass of people have chosen to believe that those worlds exist and that they matter; that the experiences possible within them are worth seeking out and caring about. If believing in a virtual world helps enhance others' experience of it, thus making that world valuable, then it naturally follows that creating experiences for others within a virtual world will help make that world valuable, too.

In an optimal metaverse, we'll be able to quantify that value. I'll explain how virtual workers will be able to profit by creating useful, fulfilling experiences for others within their worlds. This argument isn't just a utopian fantasy: It's rooted in time-tested games-industry monetization strategies. Games retain their users by providing them with fulfillment, and I believe that virtual worlds will evolve along the same principles. I don't mean to suggest that every virtual world organized along these lines will be wholly devoid of unsavory, destructive, or antisocial content and experiences. But this focus on fulfillment is at least a more promising basic monetization equation than that of today's social media platforms, which must rely on attention-grabbing tactics to provoke clicks, thus precipitating a toxic downward spiral.

As the experiences in virtual worlds become more engaging and fulfilling—again, proven in the games industry to be key to long-term retention—the people and entities providing those experiences will be able to earn more and more money. Whether you draw your salary leading a guild of dragon-slayers, or tricking out virtual cars at some phantasmagorical hot-rod shop, these jobs will be the polar opposite of the bullshit jobs that have come to dominate the real-world economy. An optimally valuable metaverse will make humane, creative jobs more accessible to more people than ever before in human history. I believe that these jobs will represent a viable solution to the crisis of purpose that affects the world of work today.

The biggest impediment to establishing a fulfillment economy is the false but persistent narrative that we should feel guilty about all the things that make us fulfilled. But what is life for if not to be maximized? And what is the point of an economy if not to improve, rather than worsen, people's lives? Productivity is not inherently virtuous, nor should it be an end in itself. Fulfillment should not be tangential to the purpose of work: Fulfillment should *be* the purpose of work. There will always be some undesirable jobs in any society, and virtual society will be no exception. But the difference in an optimal metaverse will be that the purpose of the system—of the entire metaversal economy—is to generate fulfillment. These types of dull, unpleasant jobs, therefore, are more likely to be the exception and not the norm for the majority of workers, and they will likely command a much higher premium to perform. To better understand why this is the case, let's take a closer look at the fulfillment economy.

INTO THE FULFILLMENT ECONOMY

In Chapter 5, I argued that a metaverse is a network of meaning. The worlds within an optimal metaverse will contain lots of useful, fulfilling experiences that create value for participants and for society. The purpose of engaging with the metaverse is to create and transfer this value between worlds and the individuals within them. Up until now, I've primarily discussed the concept of value and virtual spaces in terms of the psychological and social benefits generated by engagement within those spaces. As the metaverse grows, it will start to create economic value, too. If the whole point of a virtual world is to provide its users with meaningful experiences, then it stands to reason that we will eventually be able to put a dollar value on many of these experiences.

We already know that it is possible to quantify the value of a virtual experience. When people pay fifty dollars for a new video game, for example, that means the experience of playing the game is expected to be worth at least fifty dollars to them. The fact that people pay to customize their avatars in MMOs means you can put a specific dollar amount on the value they get from the experience of giving their avatar a cool new outfit. There's nothing inherently strange about people paying real money for virtual pants. If virtual experiences offer people actual fulfillment, then it also stands to reason that people will be willing to pay to have them.

As virtual worlds grow and expand, their users will find new opportunities to earn money providing goods, services, and experiences that meet other users' needs. In the real world, when you find an environment that uniquely meets your needs—a great coffee shop, a well-equipped gym—you tend to return there over and over. You tend to spend money there, too, and you tend to think that doing so is worth the expense. The same logic will apply in virtual space. As virtual worlds get better and better at meeting their users' needs, those users will come to rely on these worlds more and more, and they will not hesitate to spend money on virtual goods or services that they find valuable.

If the value provided by the worlds in a metaverse is represented by the useful experiences they offer, then the most valuable worlds will be the ones with the widest variety of high-quality experiences. But, as I've noted, it would be both financially unworkable and broadly suboptimal for the platform providers to create all these experiences in-house or pay other people to do so. Independent third parties—entrepreneurs, artists, and ordinary people—will pick up the slack. In an optimally valuable metaverse, the big economic winners won't just be the platform providers—they'll be the entities and individuals creating value by providing its users with useful, fulfilling experiences.

Employment in these worlds will be creative by its very nature. The jobs to be found in an optimally valuable metaverse have the potential to eliminate the structural inefficiencies that have made so many real-world jobs so dissatisfying for so long. The sorts of virtual jobs I'm envisioning will have very few barriers to entry. They'll be able to be performed by almost anybody, almost anywhere in the world, from a computer console or a mobile phone. There will be room for advancement within them: Someone who starts at an entry-level position will be able to build their skills and earning potential to the point where they could quite plausibly become wealthy. Human fulfillment will be the work output of these jobs, and the more fulfillment your work creates, the more money you'll be able to earn.

Unlike many potentially high-earning positions, though,

these jobs will not make the world a worse place: They won't directly harm anyone, and they won't directly exploit anyone. Unlike most jobs that begin as entry-level positions, the jobs will be creative ones right from the start. They will be forms of art where the artistic products manifest in the workers' interactions with other people. These interactions will be positive ones by design, which means that these jobs will improve the workers' lives, as well as the lives of those people with whom they come into contact.

This virtual model of labor will open up opportunities for remote knowledge work to more than just the highly educated individuals who already compete for lucrative telecommuting jobs on the global labor market. It will create a newly humane form of globalization that cannot easily be impeded by immigration control or exploited by companies that want to globalize their operations primarily so they can save money by paying workers less. Today, labor can move around, but not very easily. Virtual labor will bring about true globalization of opportunity.

In the long term, I believe that virtual society will give birth to a sustainable creative economy that taps the potential which lies inside all human beings. This transition will have transformative economic consequences for the outside world. Though not everyone will choose to do so, every person who is willing and able to exercise their creative minds will have a chance to become an economic player within an optimally valuable metaverse.

VALUE CREATION IN A VIRTUAL WORLD

The very first job within a virtual world, and thus the first step toward actuating a fulfillment-centric economy, is the job of caring about the world. When people acknowledge and care about the outcome of events in the metaverse, then the metaverse becomes made more fulfilling for everyone. The more that people care about the worlds within a network of meaning, the more valuable that network becomes. This is the first level of value creation within a metaverse.

How can an activity that doesn't actually earn money count as a job? Well, if the primary point of a job in the fulfillment economy is fulfillment, then income is not the sole and direct purpose of the job. Instead, the purpose is to confer fulfillment for the worker and create fulfillment for other people within the world. Consider massively multiplayer free-to-play games such as *Roblox*. Value is created within those games by other people showing up to play them. In many ways, the value of having so many people engaged within these worlds at all times is that there's always someone to play and interact with. While a huge percentage of the people who use today's free-to-play games don't spend any money in them at all, their presence creates value for the people who do. There would be little point in paying money to enhance your avatar's appearance within a massively multiplayer game if no one else was there to notice your new look. The fact that the world is populous makes your purchase feel worthwhile.

Sustained attention and effort within a virtual world can create value within the world, even if that value is not initially economic in nature. But as more and more people start to care about the virtual worlds they inhabit, then more and more economic opportunities will arise. That said, the real-world economy has embraced automated processes to such an extent that it might be naive to think that digital worlds—which are, in a sense, automated processes by their very nature—would buck the trend. What reason is there to believe that human labor within a virtual world would be more valuable and cost-effective than some software-based solution to the same problem? Wouldn't it just be cheaper and easier for developers to code AI programs that can use algorithms to give people fulfilling experiences?

Yes and no. AI software solutions will absolutely play a role in filling out virtual worlds and making them feel dense and populous. But, as I've established, the source of value and meaning within a virtual space comes from the presence and participation of other people. If there's no society within a virtual world, then there's no value to be found there. Your reputation as a legendary dragon-slayer in Medieval World only has value when other people within that world can sing your praises, laud you when you walk down the street, and ask you to endorse their products. (It would be much less valuable if an AI program were to slay the dragon.) The works of digital art within a virtual space grow in value because other people choose to appreciate those works. Human participation is what powers the network of meaning.

Networks of meaning are multilateral, of course; the worlds connected by them exist in conversation with one another. This linkage implies that value within a virtual world will not be a closed loop. If Medieval World becomes big enough, then eventually your reputation as a dragon-slayer will spread beyond its borders into the broader metaverse. The value you derive from that reputation will be transferable; fame and accomplishment in one world will also play out in the other worlds. YouTube stars become boxers; TikTok stars get book contracts. Is it so hard to imagine the top dragon-slayer in Medieval World sitting down with Jimmy Fallon on *The Tonight Show* to talk about their accomplishments?

Your reputation as an ace dragon-slayer grows and generates

value because other people in that world imbue that reputation with meaning. Human involvement in a virtual world is what makes the world fulfilling and contributes additional meaning to that network. It's what gives the interactions stakes and import. As we know from self-determination theory, relatedness to other people is a core component of intrinsic fulfillment. The fulfillment to be found within a virtual world comes from the choices, decisions, and actions of the individuals within the world. These choices serve to extend the reality of a world—and the more real a world feels, the more the things that happen there will matter. Human participation is the ingredient that separates a living experience with infinite possibilities from a theme-park ride that cannot depart from the confines of its track. It is what will make virtual worlds not just fun, but meaningful.

If that's where the psychological value comes from, then where does the economic value come from? Creating economic value within a virtual world, and thus creating a robust metaversal economy, begins with creating consistent earning opportunities within and between the various worlds of a metaverse. By "earning opportunities" I don't mean sporadic gig work, or the petty cash you might make on the internet from selling a chair on Craigslist and then three years later selling a bicycle. I also don't mean the sort of money someone might make by leveraging investment capital to create a platform that might one day go public. (That doesn't count as making money within a world; that counts as making money by *being* the world.) The transformative economic prospect of virtual worlds lies not in the money that a handful of companies might make by creating or hosting them, but in the money that people in the world might make on a reliable and regular basis.

There are two main ways to earn money within a virtual

world, and the first is by creating and selling virtual goods. Though the digital worlds of earlier eras are rudimentary compared to the worlds we'll soon build, the long-term data we can take from them demonstrates that an economy built around virtual goods can be as robust as one involving physical goods. For decades now, participants in virtual worlds have been very willing to pay money for goods that improve their in-world experiences and enhance their fulfillment levels.

In a 2001 paper on the burgeoning economy within the massively multiplayer game *EverQuest*—when the paper was written, the game's currency was said to be stronger than the Japanese yen, its GNP per capita the 77th highest in the world—economist Edward Castronova observed that its players were eager to spend money on a wide variety of virtual items. "These ordinary people, who seem to have become bored and frustrated by ordinary Web commerce, engage energetically and enthusiastically in avatar-based on-line markets," wrote Castronova. "Few people are willing to go Web shopping for tires for their car, but hundreds of thousands are willing to go virtual shopping for shoes for their avatar."

Castronova spent months exploring the economy of the *EverQuest* world, which he found to be fundamentally similar to our own economy. "From an economist's point of view, any distinct territory with a labor force, a gross national product, and a floating exchange rate, has an economy," he wrote. "By this standard, the new virtual worlds are absolutely real." Commercial activity took place in impromptu markets where avatars went to trade and haggle over virtual goods. Castronova's paper was interspersed with brief "diary entries" in which he recounted his own experiences participating in the *EverQuest* economy:

i made a killing in misty acorns. you can pick these up from the ground in misty thicket. i was in river-vale one day and some lady was paying 8 pp per acorn. that's a lot of money. she told me it was for halfling armor. ok, whatever. so i started making a habit of picking them up whenever i saw one, then walking into rv and selling them to rich people. they would rather spend that kind of money than wander around looking for acorns. classic economics—my comparative advantage in foraging leads to exchange. and now i can buy a nice hat.

EverQuest was no outlier. Similar economies have arisen in every other digital virtual world of sufficient size and complexity to warrant the name: *Ultima Online, Eve Online, Second Life,* and many others. (Rob Whitehead, my co-founder at Improbable, had a brief and lively career as a teenage arms dealer in *Second Life;* he used his profits from building and selling virtual weapons there in part to fund his university education.) Though admittedly these virtual economies have not had a truly substantial impact on the real-world economy, that doesn't mean virtual economies are intrinsically flimsy. It just means that these worlds were insufficiently capacious and complex to support vast numbers of users, and that there were limited mechanisms for value transfer between those worlds and our own. In the near future, these technical issues will no longer pose as much of a problem. As technology improves, so will the potential quality and value of the virtual worlds we can inhabit, and the spectrum of virtual goods we might buy or sell there.

What makes virtual goods valuable? First, they must have some utility to the purchaser within the world. While you might buy one virtual item for the sheer novelty of doing it—in the way a visitor to Las Vegas might drop a fiver on a slot machine, less for the expected return than for the experience of having done so—the initial novelty factor will quickly fade. Thereafter, if you're going to buy something in a virtual world, you'll probably want some sort of return on your investment. Consequently, the item being purchased must in some sense matter to you; it must offer some measure of utility within that world. The goods that most often do this are the ones that are novel or interesting.

For a virtual good to be valuable, it must be useful—a weapon to hunt with in *EverQuest*, a virtual house as the locus for your time in *Second Life*—or have interesting properties, preferably both. The act of imbuing these goods with unique properties is an act of creativity. Designing interesting virtual goods that people want to buy involves treating those goods like a medium for creative expression. The more expressive or unique your virtual good is, the more potential value it holds. People everywhere have always flocked to things that are one-of-a-kind.

In the real world as in digital space, value is conferred by scarcity. As Castronova put it in his paper, "Scarcity is what makes the [virtual world] so fun. The process of developing avatar capital seems to invoke exactly the same risk and reward structures in the brain that are invoked by personal development in real life . . . people seem to prefer a world *with* constraints to a world *without* them." But enforcing scarcity in digital space is a tricker problem than it is in the real world.

In the real world, most tangible goods are rivalrous, which means, basically, that two people cannot simultaneously possess or consume the same good at the same time. If you're wearing a hat in the real world, no one else in the world can be wearing the exact same hat at the exact same time. If your friend wants a similar hat, it will cost them money to acquire one, because it costs money to make the hat. If your friend wants *your* specific hat, they'll have to offer to buy it from you, snatch it off your head, or wait until you're bored with it.

Digital goods, on the other hand, can be non-rivalrous. Theoretically, the pixels and computer code that comprise a digital "hat" can be infinitely duplicated at no additional cost. With the marginal cost of producing digital goods being at zero, traditional economics says the price should be zero, too. By this logic, the only way to increase the price of a digital good is to create artificial scarcity.

That said, it's also worth noting that there exists implied scarcity within virtual worlds. In any story or any world of ideas, there are always hierarchies, importance, relative abilities and power; there's always something that makes a story interesting. There's always drama, and that drama will always imply a quantifiable scarcity in the commodities, goods, and services that make up that virtual world. If you want to live close to the palace in Medieval World, but only 1,000 single-family dwellings can fit within the palace area, well, the 1,001st person who comes along can't just build a house on top of an existing house. This person will have to pay up if they really want to live within spitting distance of the king.

These intertwined concerns of scarcity and value in virtual spaces are among the problems that non-fungible tokens are trying to solve. At the time of writing, NFTs tend to confuse a lot of people. We still primarily exist in a real-world context, and NFTs do not seem to solve any real-world problems. Much of the discourse around NFTs presumes that they are broadly frivolous instruments—and, within a strict real-world context, this characterization isn't entirely unfair. You can't blame someone for shaking their head in confusion when, for example, a basic digital image of a trash can sells for $250,000 on an NFT marketplace. (I'll hasten to add that NFTs absolutely have value as works of art or as entry tokens into digital communities.)

In virtual worlds, though, blockchain-based instruments like NFTs can be used to securitize digital assets and guarantee the unique properties of a digital good. By creating auditable interoperable rules via blockchain-based solutions, value is created as supply is restricted. In the process, it also becomes easier for people to make a reliable living producing and selling virtual goods. By imbuing a digital good with a certain immutability, as well as with the mechanisms to verify proof of ownership, that good becomes more valuable not just for the creator, but also for the purchaser.

These innovations will help turn commerce inside virtual worlds from a fun hobby into a legitimate career opportunity for many people. As I've already noted, the digital worlds of earlier generations ultimately had limited real-world economic utility, which served to curb regular users' enthusiasm for quitting their day jobs and building long-term businesses in these early virtual worlds. An optimally valuable metaverse must be constructed in a way that minimizes structural impediments to individual wealth creation.

The platform providers cannot be allowed to hoard all the opportunities to realize profit within and between virtual worlds. Creating a truly free and fair market for labor and entrepreneurship within a metaverse will be a critical part of making a metaverse into an equitable, democratic, transformative network. Building the mechanisms for storing, transferring, and securing economic value into the structure of the metaverse is a way to vouchsafe these promises and ambitions. It would be a natural and intuitive use for blockchain-based technologies, and it is a necessary step on the road to virtual jobs.

| REAL MONEY, VIRTUAL EXPERIENCES

It's ten years from now. You had a long day in the real world, and after finishing work and putting the kids to bed you want nothing more than to be around other people and relax in a friendly, familiar environment. So you strap on your headset, log into the metaverse, and head over to Roger's, your favorite virtual tavern in your favorite virtual world: 1990s World. The benefits of virtual pubs are obvious: You'll get the same convivial atmosphere you'd get in a real pub without having to actually travel to the pub, pay for a babysitter, or wake up the next day with a hangover. But the specific selling point of Roger's is Roger himself: the warmhearted, flannel-clad bartender who's always quick with a joke, a sympathetic ear, and a story about that one time he met Kurt Cobain.

Roger's always there because, well, you pay him to be there, you and all of the other customers who have come to value his companionship. From 5 P.M. to 2 A.M. six nights per week, you can count on Roger being behind the bar to pour drinks, tell stories, and broker introductions between customers. Running this virtual tavern is Roger's real job, one that plays to his strengths and competencies as a raconteur and a connector of people. It's how he earns his living, and as he welcomes you inside and slides a pint down the bar toward your standard stool, you smile and send him a payment in 1990sCoin—the in-world cryptocurrency for 1990s World—knowing that being here is worth every cent you pay for the experience.

You're willing to pay to keep Roger's in business because the place means something to you, even though it's an entirely virtual space (and even though you have to buy your own real-world beer). Likewise, you're willing to pay Roger himself to show up six nights per week because it's worthwhile for you to know that your favorite virtual bar keeps regular hours, and that you can count on your favorite bartender to be there just in case you want to drop by. This value proposition exemplifies the second major way that individuals can make money in a virtual world: by creating and producing fun and useful experiences for other participants in these worlds.

The value in an experience such as hanging out at Roger's cannot be derived from automated functions alone. It is also a product of the actions of people within the world. While it takes programming work to make virtual worlds feel rich and immersive, it will take human labor within the rendered world to provide the experiences that make these worlds feel sophisticated and present.

Why can't the experiences in virtual worlds work like they do in advanced video games, where deep programming allows players to explore open worlds inhabited by other characters, many of which can interact with you in one way or another? While NPC (non-player character) interactions might be engaging over the short term, they are not very fulfilling over the long term, which is why virtual worlds must pick up where video games leave off. Though its constituent worlds might share certain surface-level similarities with video games, the metaverse is not a video game. It's not a solo activity within a finite system. It's a network of worlds in which value is tied both to the number of participants and to the fulfillment levels that those participants can experience for themselves and create for other people.

The most valuable and popular games in the world today, by a huge margin, are multiplayer experiences, which very obviously illustrates the value proposition at work here. The presence of other real people within digital space—beings whom you acknowledge to be part of your actual society, rather than just humanoid figures operating as props—is the basis of how the metaverse generates fulfillment. Is it exciting to watch a football match with thousands of NPCs who exist only to fill out the crowd? The issue is not one of verisimilitude—one can imagine AI-powered NPCs becoming very real, indeed—but one of meaning. It just wouldn't mean as much to see excitement on the faces of characters whose entire purpose is this one event: beings with no life beyond the bleachers and no relationship to society at large.

Let's say you could make another world in which you were God, and let's say that all of the beings in that world—beings that, no matter their genesis, were all *real people* within the parameters of that world—did everything you told them to do. Next, let's imagine that you unplug from that world and come back into your real life, only to be confronted with the depressing fact that nothing that happens in the other world matters in the real world. This imbalance is not a setup for fulfillment, and the two worlds in the set certainly don't comprise any sort of valuable network of meaning. The beings that you control in the other world might look, talk, act, and seem wholly real within that world. They might be intelligent, realistic beings with rich inner lives, the products of an AI so advanced that they are functionally indistinguishable from you and me. But their lack of connection to the rest of your existence and to a wider network of worlds limits their ability to generate value within a metaverse.

We've long known that human labor and interactions can add value to online experiences in ways that cannot yet be matched by mechanized interactions. The moderator of a group or a comment section, for instance, adds immense value to their forum by deploying their human judgment and intelligence to keep discussions focused, productive, and respectful. Another instance of human labor adding value to a virtual experience is the gaming phenomenon of "boosters": skilled players who will play a game with you, or for you, in order to boost up your ranking. "Boosting" isn't necessarily a form of cheating or of skipping the line. Sometimes, skilled players can have real-world obligations that may keep them out of a game for some time. In that case, they might hire a booster of equivalent skill to play on their behalf, so that their ranking doesn't slip during their absence, and so that they can pick up where they left off when they finally return. At other times, novice players might want to join a game exclusively in order to play with their friends who have been there longer than they have; these sorts of people might hire a booster to rapidly advance them through the game up to the point where they can join their buddies and thus get what they need out of the experience. Boosters are the digital equivalent of trail guides, helping people reach their desired destinations even when they might not be able to get there on their own.

More broadly, stories created by a human are often fulfilling expressly because of the social context in which they are told. The ongoing popularity of the game *Dungeons & Dragons* is a great example of how human storytelling within a group setting can help turn a game into a world. Though analog in its execution, *D&D*'s human storytelling element makes the game as complex as any carefully rendered digital world. Operating from a set of universally understood parameters—the rulebooks and the dice—the game's participants, under the supervision of an improvisatory narrator/referee called the Dungeon Master, create their own characters and figure out the game as they go. Those characters grow and change over the course of a *Dungeons & Dragons* campaign. Relationships form, storylines develop; players solve problems together and share in defeats. The open-ended nature of *D&D* means that a campaign can plausibly last for years, and some of them do. Rather than create a closed world with predetermined obstacles that increased in difficulty as the game went on, culminating in a final challenge that rendered the player either a winner or a loser, *D&D* creators Gary Gygax and Dave Arneson built an open world where participants can sketch the details of the game they wish to play.

These are just a few examples of how human interactions within ingenious contexts can create intrinsic value for the users of a digital space or virtual world. Human involvement will be immensely important within the worlds of a metaverse. If such worlds are machines where fulfillment is the output, then human creative labor is what will keep those output levels high. This sort of labor can also play out in the service of exciting experiences, not just prosaic ones like sitting in a digital tavern.

Imagine that you've joined Heist World, a virtual world designed to give its users the experience of planning and executing a complicated heist like you might see in the movies. There are lots of factors you'd need in order to make a virtual heist an exciting and fulfilling experience, not least being the element of danger: the prospect of getting caught by the police. What fun would it be if you knew for a fact that you would succeed and get away scot-free? That's not a challenge, that's a cheat code. For Heist World to feel like a world and not just a game, the police can't be NPCs whose movements you can predict and whose attention you can avoid by keeping to the far left side of the screen. Your experience as a master thief in Heist World will be enhanced if you're matched against other users who assume the role of police; if you must outwit and interact with real humans, not just a program.

If you accept that premise, then you can also understand why a thief in Heist World might be willing to pay money to hire people to clock in as virtual police. The presence of a reliable, dedicated squad of detectives would make for a better, more fulfilling experience for you and your team. Likewise, you can also see how a virtual police officer in Heist World might want to pay you to plan new heists each week, because having a reliable human antagonist within the world improves their experience of that same world. If you can picture thousands and thousands of people caring enough about a virtual world to want to pay to make it better for themselves and others, then you can picture an economy springing up around these people: virtual police, virtual thieves, virtual bartenders, virtual everything. You can see the thieves in Heist World logging in to 1990s World to plan their heist at Roger's. You can see the cops tracking them there. You can see the ways in which these disparate worlds start to interrelate.

The virtual worlds of earlier eras were limited by the technologies that powered them, which could render worlds of only limited capacity and depth. But the technology that powers today's virtual worlds is better than ever before, which means that more and more people can do more and more things within them. As these worlds grow in complexity and scale, and as demand increases, these informal jobs will become formal jobs. For example, eventually it will really matter to the patrons of Roger's that the bar is open when they want to use it, and that Roger is reliably there and available to take their orders and dispense drinks and companionship. Pretty soon it won't be enough for Roger to show up only when he feels like it. If the product he's providing is sufficiently valuable to enough participants, the economic incentives will be such that he'll feel compelled to be "on duty" on a regular schedule, just like with a real job—but with many advantages over a real job, not least because he can do the work from the comfort of his favorite chair at home. The job won't involve the tedious physical labor of being a bartender, such as changing the kegs, dealing with aggressive drunks, or having to stand for ten hours per day. Instead, it'll focus wholly on creating value by providing engaging experiences.

I've focused thus far on "artisan" labor within virtual worlds, goods and experiences that mostly involve person-to-person transactions and interactions. But virtual worlds will also create a vast market for more complex experiences: quests and adventures and festivals and fantastical occurrences that will require significant resources to create and execute. These will likely be the province of businesses more than individuals: independent entities that might be able to spend tens of millions of dollars programming some incredibly immersive and detailed space adventure, for instance, or creating and convening a massive festival. These will be legitimate businesses built within virtual worlds, and there will inevitably be jobs to be had within them.

You can think of the spectrum of opportunities that will be available in virtual worlds as a ladder of sorts. The baseline assumption is that people will spend time in a virtual world; with that time, they will perform labor that makes the virtual world more valuable for other users. But a novice won't immediately have advanced skills to offer or be able to create complicated, high-value experiences. These beginners will therefore start out at the bottom of the ladder, pursuing the sorts of jobs and opportunities that fit their skill sets. As their skills grow and they ascend the ladder, the world will provide opportunities for more complicated work—perhaps the opportunity to create experiences for other people; perhaps the opportunity to play a meaningful role in one of the grand experiences referenced above.

One big reason why virtual jobs haven't yet evolved into a significant economic force is that there haven't been enough strong links between these virtual worlds and the real world—or links among different virtual worlds—which means there are fewer opportunities to transfer value between them. Creating a strong "meta" layer between virtual worlds and the real world will make it more and more plausible that "virtual jobs" will evolve into actual primary means of employment for a non-negligible percentage of participants. So, how do we continue to optimize for the worker's fulfillment when a virtual job becomes an actual job? And how do we make it so that the virtual economy makes capitalism better, rather than just replicating its existing flaws in a different sphere?

A NEW PARADIGM FOR VALUE

The question isn't whether or not the virtualization of work *should* happen. It *is* happening already, and it will continue to happen. But we are still at the point where we can write the future of virtual work. We have a chance to reinvent the orienting principles of employment such that virtual work avoids many of the pitfalls of the current working world. We can pursue a model that creates new jobs localized in the context of virtual worlds rather than just extending the old, broken work structures into a virtual context. This model is the one we must follow if the virtual economy is to improve upon our current one.

Although it might seem like a long road from here to there, the fact is that many of us are already working virtual jobs. The COVID-19 pandemic accelerated the trend toward jobs entirely mediated by computer screens. Thanks to remote teleconferencing apps, workflow software, social messaging services, and the rise of "work from anywhere" policies, many of us never meet our co-workers in person, never see the inside of an actual office, and have no direct connection to our work products.

There are surely lots of people reading this book right now who spent years longing for the day when they'd never again have to commute to the office or hobnob at the water cooler with their co-workers. So now that this dream is finally starting to come true, why does it feel so unfulfilling? For many workers today, the virtualization of work has been an unpleasant experience. I would suggest that this ennui is in part the result of employers unsuccessfully trying to reconcile a new medium with an old context. Meetings are bad enough in the actual office, and it makes very little sense that the virtualization of work has led to more meetings rather than fewer. The effacement of the lines between home and work has made it feel like our workdays never end, which is frustrating precisely because the workday used to end when we left the office. We have taken the tools that could be used to generate fulfillment and have used them to try to boost production levels by turning the home into the office. This is not the path to take if we wish to reinvent work for the era of virtual society.

Likewise, many of us are also already accustomed to working for companies that produce what could plausibly be described as virtual goods. The data economy, as I mentioned, does not primarily produce items that you can hold in your hands. Google, Facebook, software companies, games companies, online content providers: Their work products are also exclusively mediated by screens. These are all, arguably, virtual companies that produce virtual products. The difference is that these virtual companies, many of them, are still based in a real-world context, which can create cognitive dissonance and make the jobs there feel in some respects suboptimal ones.

Today's internet-centric virtual jobs are largely focused on enhancing real-world commerce in some way: a buyer finding a seller, a community forming around a brand, news or entertainment content rooted in the real world. The difference between these sorts of virtual jobs and the ones we will create in the metaverse is that the point of the metaversal economy is not to improve efficiencies for real-world companies. The enterprises that will thrive in the metaverse will be rooted in fundamentally different premises than those that animate real-world commerce. Their goals will be to produce fulfillment, rather than goods or data. And I believe they will precipitate the same sort of seismic macroeconomic shift that we last saw around the turn of the century.

At the end of the twentieth century, the global economy was still stuck in the past. In 1996, General Motors sat atop the Fortune 500, in the same spot it had occupied when the list was inaugurated in 1955. Close behind General Motors on the list were several other old corporate stalwarts: Ford Motor Company, ExxonMobil, AT&T—all of them companies whose business models (cars, oil, telephones) your grandparents would have understood. For the most part, the massive corporations that ran the world through the end of the twentieth century presided over an economy of tangible things.

By 2021, though, both the world and the global economy looked a lot different than they had a quarter century earlier. While the top slot on the Fortune 500 in 2021 was occupied by Walmart, slots two and three were occupied by Apple and Amazon, respectively, with Alphabet, the parent company of Google, not too far behind. (GM, Ford, AT&T, and ExxonMobil occupied slots twenty-two, twenty-one, eleven, and ten, respectively.) The world's biggest companies by market cap were Apple, Microsoft, Alphabet, and Amazon, with Tesla and Meta grappling for the fifth spot—all six of them trillion-dollar companies, five of them paragons of the new data economy.

Within the span of twenty-five years, the transition from a physical economy to a data economy had created some of the biggest companies in world history. This growth trajectory was catalyzed by the emergence of the internet. The network had created a suite of new opportunities and needs, and the people who best understood how to fill them were able to profit handsomely. The big companies that had dominated the world in 1996 were mostly still around in 2021—but, given that they were inextricably rooted in yesterday's economic priorities, they were simply less important to the present and the future. The rise of the internet had created a new paradigm for value.

The rise of the metaverse will present a similar paradigm shift. In the near term, the economic opportunities and value inside virtual worlds and the metaverse will be as big a disruption to the real world as was the economic value created by the internet. Just as the internet turned digital data into a precious resource to be organized and searched and displayed, the metaverse will commodify the sorts of useful, fulfilling experiences that humans have always sought from their constructed social realities. Just as the internet offered new and fertile ground from which new types of businesses could emerge, digitally rendered virtual worlds will give companies and entrepreneurs new contexts in which to extend their economic ambitions.

As these worlds scale, I believe that millions of workers may well end up earning their primary incomes within them. Unfortunately, the legal structure of the real world hasn't yet caught up with the many questions that will arise once the metaversal economy takes flight. When government bodies get around to realizing that a new technology has created a social problem, it is often already too late to solve that problem. Today, the massive trillion-dollar companies that dominate the data economy aren't just too big to fail: They're arguably too big to be fixed. In terms of the influence they wield and the autonomy with which they wield it, companies such as Google and Facebook are, effectively, their own countries. World governments now engage with these companies less on a supervisory basis than on a diplomatic basis.

We must bear this dynamic in mind as we look to devise a capable and effective regulatory framework for the metaverse. Since government regulation tends to lag far behind the pace of technological change, a satisfactory framework will have to involve some hybrid of the real world getting involved in metaversal affairs and the growth of a truly robust governance structure within the metaverse itself, perhaps nurtured by the Exchange. Once these worlds become important enough that it feels natural for them to engage in self-governance, we'll start to see new paradigms emerge for politics, democracy, and freedom. This will be the point where virtual worlds truly become a virtual society. But what comes after that? How do we create a regulatory superstructure to ensure that virtual society doesn't replicate the problems found in real-world society, real-world capitalism, and the internet? I'll address these questions and more in the next chapter.

Chapter 8

THE TYRANTS
AND THE COMMONS

W hen the definitive history of the internet is finally written, it is likely that our present moment will be deemed a digital dark age. Much of the internet today is dominated by massive, amoral companies that profit by gathering and exploiting user data. These companies and their network effects are so large that it is difficult for competitors to challenge their dominance. Considering the influence they wield over users, content creators, and entrepreneurs, they have become the unelected imperial governments of the new world.

For all the talk of startups or disruptive companies in Silicon Valley, the most powerful entities in tech aren't really companies anymore, or even really monopolies. They are global empires without conventional militaries. Google's roughly $1.8 trillion market capitalization is higher than the GDP of Russia—high enough, in fact, to rank the company as the eleventh most prosperous nation in the world. As of Q4 2021, Meta reported 2.82 billion daily active users of Facebook, Instagram, Messenger, and/or WhatsApp: more than the population of China and India combined.

Instead of conquering new lands, the emperors of the internet have settled for reducing their already vast populations—their users—into a state of digital vassalage. Without the billions of posts, photos, likes, groups, and comments added and created by its users, Facebook would be just another website. Without the countless individuals creating videos in pursuit of self-expression and perhaps a lucrative viral hit, YouTube would be just another defunct video store. And yet users retain only a fraction of the value they create for and within these platforms.

As a gross simplification, you can think of today's big tech platforms as having two major components. The first is the "fun part," or the user-facing features: the photo filters and "like" buttons and instant messages. The second component is a gigantic proprietary database that represents the value of the network. The fun parts of the platforms are what attract users. The massive living database is what makes these platforms their money.

Facebook's main products, Google Search, Amazon, Tencent's WeChat: All of these services are powered by big databases controlled by these companies. These databases—of users, or websites, or messages—make these companies utterly unstoppable in their core businesses. The act of using their services generates an ever-expanding corpus of data that they are then free to mine and exploit to their own tremendous advantage. They reap limitless riches by gatekeeping the information that represents the very structure of our world.

If you are reading this book, I suspect you are quite familiar with this line of thinking. But you might not recognize that there isn't necessarily any malice in this whole process. Early-stage tech businesses are motivated primarily by frantic survival, and then by the modern investment mandate of growth at all costs. Once they start to grow, they are no longer designed: They just happen. One tech company founder once confided to me that after hitting a certain level of "product market fit," none of his decisions mattered nearly as much as they did in the beginning: Good, bad, the company simply grew. Network effects can make this growth trajectory seem inevitable.

But rhetorical justifications for corporate growth are often self-serving. Massive companies have always proclaimed their own inevitability as a means of dissuading competition and discouraging outside scrutiny. Government regulation is supposed to constrain corporate overreach and provide a framework for socially responsible growth. For centuries, antitrust law has been deployed to break up business monopolies. Modern-day regulatory bodies, such as the Food and Drug Administration in the United States, help to ensure that certain products are not rushed to market before they've been shown to be safe and effective. Looking after the public welfare by imposing certain limits is one of the fundamental roles of government in a capitalist system. Even Milton Friedman, in *Capitalism and Freedom,* which took an extremely free-market view of economics, conceded the important role of regulation in protecting society from the negative externalities of business in areas such as health and the environment. But these regulatory structures, so effective in other arenas (when considering the developed world at large), have mostly failed to rein in the internet behemoths or temper their ambitions.

Over the past twenty-five years, as startups moved fast, broke things, and achieved billion- and trillion-dollar valuations, regulatory bodies were slow to recognize exactly how the economy was changing. Most lawmakers did not realize what made these companies different from other tech businesses, or why they perhaps should not be allowed to wield sole and total control over the vast stores of user data they were accumulating. Facebook's Mark Zuckerberg wasn't asked to testify in front of the U.S. Congress until 2018—fourteen years after his company was founded, when the company was by then valued at nearly half a trillion dollars—and the ensuing hearing served mainly to illuminate just how little most legislators understood about the internet. "How do you sustain a business model in which users don't pay for your service?" octogenarian Republican senator Orrin Hatch asked Zuckerberg during the hearing—a question that makes sense only if you presume that Facebook is just like any other business.

Massive internet companies aren't just businesses providing a service, and the problem they pose to society isn't merely that, like the monopolists of the Gilded Age, their size and efficiencies inhibit competition. These companies are monopolists of the new digital commons. A *commons,* broadly, is a public resource that serves the general welfare; the commons I reference here is, specifically, the vast combined mass of user data that these companies control. The material within the commons, powered by the contributions of innumerable people, is usually created for free. The data commons that has emerged from these platforms is as critical to the functioning of our present and future society as clean air and water, national defense, and built infrastructure. These are all classic public goods over which governments have typically asserted control. And just as government can decide how fast people are allowed to drive on public roads, and can stop companies from dumping waste in the river just because they'd rather not pay to have it hauled away, so too is it absolutely the business of government to decide how that commons is used: who controls it, who can access it, whether it can or should be allowed to enrich private entities, and how it might be used for the benefit of all. This data represents present-day society in its purest state. It represents reality for the people who created and interact with it, and it defines the virtual societies for which it was created.

The biggest internet platforms serve to extend our social realities into new frontiers, but the people who control those platforms are, at present, free to unilaterally dictate the parameters of these new realities. If these platforms can be said to represent a stage in the evolution of virtual society, then the stage they represent is tyranny. These platforms' constituents, after all, are thoroughly disempowered: They cannot exert pressure on these platforms or participate in the mechanisms of governance. The workings of these platforms are almost completely opaque. Users can earn money within them, but only under strict guidelines, and the platforms have the power to destroy a user's business at any time, for any reason. Because these companies so thoroughly control the data that resides within their commons, users cannot leave these platforms without sacrificing some meaningful part of their digital identities.

Perhaps these companies somehow benefit us by wielding continued power over the commons? Perhaps they use that power to create ongoing value that otherwise would not exist? The argument that these are highly innovative companies, deserving of privilege for the benefit of all, is quickly demolished through even a cursory financial analysis. These companies spend wastefully, sometimes dropping billions of dollars on fanciful boondoggles. The extraordinary profits of their core monopolies often disguise a tremendous inability to build new products instead of acquiring them. These are signs of a lack of healthy competition.

Even though some of these companies are publicly held, their equity structures are often set up to disempower regular shareholders. Their founders, often, are accountable to no one but their biggest shareholders and their C-suites. While it would be hypocritical of me to argue against founder control in all instances—it's a privilege I myself enjoy—it's hard to justify total founder control when entire countries might run on a given platform without having much choice in the matter. History is littered with examples where absolute power combined with zero accountability has led to terrible outcomes. If nothing else, the consolidation of corporate power into the hands of a few tech founders could return us to a world of monarchic succession crises. Even the best, brightest, most deserving founder can suddenly die. Where, then, does that leave the billions or even trillions of dollars of economic activity dependent on their unilaterally controlled platforms?

Governments haven't just been slow to regulate the new tech behemoths: They have struggled to effectively understand the scope and import of the regulatory challenge in front of them. If regulators insist on viewing wholly new businesses through outdated frames of reference, they will always misunderstand them, and their regulatory efforts will fail. In the absence of intelligent regulation, the companies that emerged to dominate the internet were left alone to govern themselves—which they didn't. Now these companies play an outsize role in determining how culture, society, and democracy evolve all over the world. We've seen how that has turned out.

In our current digital dark age, bizarre social and political theories abound and spread faster than they ever did in a non-networked world, thus destabilizing polities and empowering demagogues everywhere. Abusive behavior thrives on social platforms and, far from being inhibited by being tied to one's real-world identity, is in many ways emboldened by the connection. Platform-specific attempts to counteract or neutralize this bad behavior always feel both arbitrary and futile. Discourse on the internet is doomed because the network was broken to begin with. Thirty years ago, idealists believed that networking the globe would bring about a renaissance of learning, communication, and democratic engagement. It also brought us Facebook and QAnon, a weird conspiracy doom cult that believes variously in the return of John F. Kennedy, Jr., and the pseudo-divinity of Donald Trump.

The worlds of a metaverse will matter more and feel more real to their users than the internet ever did. These worlds will create new frontiers within which to be human—which, if we're not careful, also means that they might become spaces within which to be our worst selves. If you think QAnon and its associated conspiracyverses are bad enough now as a product of memes and message-board posts, just imagine if QAnon were the basis for an actual three-dimensional world in which people could live. While a valuable metaverse has the potential to confer unprecedented fulfillment, a malignant metaverse could do real psychological damage that would create negative value for the real world.

Today's digital oligarchs and other bad actors will assuredly look to transpose the lopsided value proposition of the modern internet onto the metaverse, maximizing their own advantage by disempowering their users. It is not hard to see a trillion-dollar global economy appearing inside the metaverse. By providing services and experiences to the metaversal individual, the metaverse will grow into a space that will rival the real world in its importance and centrality to its users' lives. In the absence of a capacious framework for regulation, the companies that effectively became autonomous nation-states in the current stage of the internet will extend their control into the coming era of virtual society, a maneuver which would make the metaverse better and more profitable for them, but worse for everyone else. It would be folly to leave the governance and supervision of this prospective future to the companies that have abdicated these responsibilities thus far.

We mustn't pretend that we can pause evolution, transform human nature, or wholly constrain corporate growth imperatives. But we also mustn't become defeatist and presume that it is impossible to encourage and incentivize positive activity within and throughout a metaverse. Though we might be living in a digital dark age at the moment, it is worth remembering that, in world history, the so-called "Dark Ages" (the era is now thought unlikely to have been quite as dark as the popular imagination presumes) were followed by the Renaissance. We have volition. We can choose to maximize the promise of the metaversal age by implementing certain structural conditions that can nourish, rather than constrain, human potential. An optimally valuable metaverse is ours to create. In order to do so, we first must learn from and respond to the mistakes of the internet.

THE DOWNSIDE OF DECENTRALIZATION

The problem with the modern internet is often mistakenly described as a problem of misinformation. While misinformation is indeed endemic online, and while the real world has suffered tangible consequences as a result of its spread, I see it as more of a symptom of structural failures than an underlying disease. Regulatory lapses aside, the main structural problem with the internet is rooted in the network's naive architecture, which did not anticipate the adversarial nature of a decentralized network's economic incentives. A decentralized internet was supposed to be the wellspring of hope for a better future for humanity. Instead, for efficiency's sake, decentralization led to complete centralization.

The nature of any service that becomes popular on the internet, including cryptocurrency, is to drift toward centralization. In *The Square and the Tower*, the historian Niall Ferguson talks lucidly about the origins of this problem, citing examples throughout history to show the pattern repeating. In summary, networks crave efficiency. Like water following old tracks, networks always center around important nodes. It is more convenient to go to one place and to conduct all of your transactions with one party than it is to go to forty places and transact with forty parties. The power of economies of scale multiplies in the world of information, which is unbounded by the limits of physical space. Today, a good idea can have such profound gravity that half the world can adopt it in a few years.

What this means in practice is that, far from breaking up the old power structures in society, successful tech companies often grow so powerful that they become the de facto centers of society. They begin as the new *agora* and eventually become the new *polis,* before morphing into the seats of empires. But while real-world emperors have certain checks on their power, and must be in some way responsive to the needs of the people whom they govern, modern social media companies are beholden to no one. The nature of the internet doesn't just create centralization, it demolishes any need for real cooperation and alliances. The biggest tech companies exist above the government and above the gravitational pull of public opinion. Eventually, these companies stop needing the consent of the people whom they nominally serve.

This popular disempowerment is exactly the problem that network decentralization was supposed to solve. In the industrialized brick-and-mortar world, the production and distribution of knowledge and information were centralized, dominated by a relatively small number of publishers and broadcasters that shaped public conversation. The people who ran the newspaper decided what was published in the newspaper, in other words. As a reader of the newspaper, the best you could hope for in terms of input was maybe to have a letter to the editor published every now and then. You had to pay to read the newspaper, too, and if you stopped paying for your subscription, then you stopped getting the newspaper. This centralized model put immense power and influence in the hands of the people who controlled and represented the production apparatuses—publishers, network executives, prominent journalists—while making everyone else mere passive recipients of news and knowledge. This model also functionally excluded any ideas or viewpoints that fell outside the sphere of mainstream consensus.

Decentralization brought about by the internet sought to invert this power structure. In a decentralized model, central control over production and distribution is shattered, and power instead goes to the individual nodes. Data on the internet is stored and accessed on countless individual computers all over the world rather than being consolidated onto one gigantic mainframe. This structure makes the internet resilient and redundant, because the health of the network is not contingent on any one individual node; it also theoretically makes it difficult for any single entity to completely regulate access to the material available online. A determined user can always find a way around a paywall, which demonstrates just how hard it is to levy tolls on the information superhighway.

The idealists who populated the early internet had hoped that this diffuse power structure would create newly egalitarian methods of communication and exchange. Decades later, though, it is clear that decentralization instead replicated real-world power disparities and control mechanisms. Looking back at the origins of the internet from our current vantage, it is easy to spot some of the failure points.

The network's decentralized architecture is content-neutral, meaning that there are no feedback mechanisms by which to either incentivize useful content and ethical behavior or prevent misuse of the network. The network itself observes no difference between, for example, a website that makes it easier for people to locate and sign up for COVID-19 vaccine appointments and a website arguing that COVID-19 vaccines are a deep-state ploy to inject us all with tiny tracking devices.

You could argue that this neutrality is a good thing, insofar as a content-neutral platform allows for true freedom of speech. But while freedom of expression is a real and valid social prerogative, in the real world it is also important to reward valuable contributions to the discourse while disincentivizing fractious, antisocial behavior. Online, though, in the absence of articulated standards for ethical use, the entire concept of "misuse" is reduced to a shifting cultural norm that is defined differently depending upon the online community in which you find yourself.

Within and between online communities, there are no universal infrastructural methods by which individuals can secure, control, and develop their own identities. Consistency of identity is an afterthought on the internet. In a world where you have to create two hundred different user profiles for two hundred different websites, few of which offer you any meaningful opportunities for self-determination, there is little incentive to invest in or feel protective of any of the myriad digital personas that you're forced to maintain.

Antisocial behavior online proceeds directly from these weak notions of identity. Like individuals who cannot recognize themselves in a mirror, I would argue, many people find it hard to truly see themselves in their various online identities, and this blinkered vision also impedes the development of empathy for others. Because we lack real autonomy over our social identities on massive internet platforms, those identities often become crude depictions of our true, complex personalities; they become masks that conceal our true selves from the world. Internet trolls and other malignant denizens of the internet act out in part because they cannot envision or perceive the humanity of others online, and in part because there is very little humanity within their own identities online. With no clear rewards for good citizenship, and no innate incentives for evolving your online identity into another frontier of your inner self, it is often easier and more rewarding to just be bad.

There are no clearly defined penalties for flooding the network with junk and misbehavior, and there are also few infrastructural methods for removing and discouraging junk. The only entities that can take concrete steps toward removing junk online are the platforms that also profit by its distribution. This dichotomy is irreconcilable. The network itself can neither disincentivize junk nor differentiate between useful and harmful content, and the companies that dominate the network have few incentives to do so. It is easy to get people to become addicted to junk.

Though idealistic in many ways, the people who organized the internet didn't build any mechanisms for ethicality into its infrastructure. As a result, unconstrained corporate capture and the ensuing disempowerment of users have made it into a fundamentally unethical space. In the absence of any structures to incentivize useful behavior and to ensure and perpetuate community governance of the network, the largest and best-capitalized actors on the network have stepped in and assumed de facto control. The digital utopians who laid the foundations for today's internet believed that the network would present a brighter, better future for humanity. But transformative technology alone is never enough to change users' ingrained patterns of behavior.

WHY THE SPAMMERS (ALMOST) ALWAYS WIN

The notion that new communications technologies will automatically lead to enlightenment is a recurrent fallacy. Innovators always want to believe that everyone else is like them, and that the users of a new thing will abide by the creator's vision and code of ethics. As Niall Ferguson has observed, the introduction of movable type in Europe in the sixteenth century was initially met by the hope that it would inspire a new age of religious devotion. By empowering people to read the Bible in their own language, the printing press was expected to inspire a renaissance of belief prompted by a technology that made holy texts more accessible than ever before.

While literate Europeans were indeed excited to read the Bible in the vernacular, by no means was that their only use for the printing press. As Ferguson notes, one of the first "best sellers" was the *Malleus Maleficarum,* a book that showed people how and why to identify and destroy the witches concealed among them. Give people the opportunity to do something holy and they will inevitably end up doing something human.

When World Wide Web creator Tim Berners-Lee, influenced by the philosophy of free-software advocate Richard Stallman, decided to disclaim ownership over his software and instead make it free to all, he hoped that the Web might precipitate a new golden age of learning and free expression. There were no real centralized, codified rules or instructions for how to use or contribute to the internet, which at first made it fertile ground for creativity and ingenuity. People from all sorts of different backgrounds could publish websites and contribute to projects and be judged primarily on the quality of their work rather than on their backgrounds or credentials. But the absence of gatekeepers also meant that there was no easy way to ward off the barbarians when they inevitably arrived.

The anarchic tendencies of the early internet were at first constrained by the self-selecting nature of the participants and the values and outlooks they inevitably shared. Many of the early hackers and network adopters came from similar academic or countercultural backgrounds, which informed their belief that the network might be the new Library of Alexandria, a vast store of knowledge from which everyone could benefit and to which everyone could contribute. Many hoped that it would bring about new paradigms of collaboration and expression, and that, rather than being constrained by mainstream standards, people could devise their own methods of doing things. Can you fault these well-intentioned early adopters for assuming that those who followed them online would act in similarly good faith?

In 1994, as Finn Brunton recounts in *Spam: A Shadow History of the Internet,* the first commercial spam message was sent to thousands of Usenet newsgroups by two attorneys advertising their services in helping non-U.S. citizens obtain green cards. These attorneys, Laurence Canter and Martha Siegel, realized that while the network wasn't meant for commercial activity, neither did it present any structural impediments to commercial activity. The worst consequence that Canter and Siegel would face would be a bunch of people yelling at them online; the prospect of commercial gain outweighed any fears of making people angry. (If this was their calculation, it was not entirely accurate over the long term; in 1997, Canter was disbarred in Tennessee, in part because of his online advertising methods.) Their second effort at spamming Usenet was met by the first instance of trying to combat commercial spam; an annoyed computer programmer wrote some code that automatically deleted the messages from the groups. It was a noble albeit futile effort, insofar as it showed that the best outcome the digital utopians could ever hope for in their quest to keep the internet pure would be an endless war of attrition against those who hoped to exploit it for personal gain.

Was it unethical for Canter and Siegel to spam Usenet? On the one hand, the people who used these newsgroups clearly did not want their newsreaders clogged with irrelevant content. The spam messages were sent without much regard for the norms or desires of the communities, which were not accustomed to having commercials intrude upon their discussions. The spam seemed unethical because it transgressed community norms in pursuit of individual profit.

But the norms were, in the end, only norms. They weren't codified, they weren't compulsory, and they weren't coded into the structure of Usenet. The norms were effective only if every single user freely agreed to abide by them, and as soon as one actor chose not to abide by them, they fell apart. Is it unethical to look out for one's own self-interest within a system where ethicality is an ad hoc social construct? Such behavior is inevitable within a naive system.

With no infrastructural mechanisms for removing or punishing those who transgressed the network's founding philosophies, no way to constrain the pursuit of financial advantage on the network, and no clear way to incentivize and reward useful, prosocial behavior, the internet eventually fell into the hands of those who wanted to accumulate and exploit data rather than refine it, improve it, and share it with the world. Many of the exploiters spoke the same language as the utopians, promising that their products would improve the world by connecting people and making it easier than ever for them to find, share, and react to useful and interesting content. Initially, at least in part, many of these companies lived up to their promises.

Every website and internet application that has since taken a darker turn was useful and interesting once upon a time. Google did offer a better way of organizing and searching the information available on the internet. Facebook did make it easier for far-flung individuals to resume old acquaintances and forge new ones. These companies and others created value in ways that users appreciate, insofar as they improved functionality while reducing or eliminating the up-front costs of participation. But the profit and growth imperatives animating these companies meant that they would always make choices that first served their own interests, rather than the interests of the community.

As we've discussed, the root of these companies' hold on the internet is their ownership over vast databases of useful information—users, things for sale, messages, and more—to which they ultimately control access. Ironically, the "decentralized" vision for the internet has been replaced with a few highly centralized systems. Part of why this happened is because the alternative—decentralized databases powering utilities shared by many companies and individuals—has numerous challenges that have previously made it impractical. One of the biggest challenges involves dealing with bad actors operating over some collectively owned commons. Decentralized systems struggle to compel individual nodes to agree to do what is best for the entire network. In computer science, a foundational concept relevant to this issue is known as the Byzantine generals' problem.

The Byzantine generals' problem, first posed by Leslie Lamport, Robert Shostak, and Marshall Pease in 1982, is rooted in an allegory. In ancient Byzantium, there were several generals of the same army, camped separately outside an enemy city, who needed to agree on a course of action. Despite their separate encampments, the generals had to find ways to communicate so that they could decide on an optimal attack plan. But none of the generals could completely trust any of the other generals not to be traitors, and none could trust that a traitorous general wouldn't provide false information that would lead the group to make bad decisions. The problem, then, was how to structure their communications so that the loyal generals could all agree on the same plan, while eliminating the risk that the traitorous generals would lead them toward a bad plan. How do you optimize for network success in a system where you know that, at a certain point, some nodes will fail the group? Is it possible to build a system that is strong and resilient enough to produce positive outcomes while withstanding individual actors' efforts to gain advantage for themselves—a system that can sustain multiple betrayals without completely collapsing?

A Byzantine fault-tolerant system is one in which all actors agree on a common goal or strategy even as they also realize that some of those actors will prove themselves to be unreliable and self-interested. This sort of system is built in a way that can withstand perfidy and unreliability among its components. While there are many approaches to solving this problem, a particularly relevant one to the metaverse is the use of blockchain-like technologies.

By logging transactions in a public, distributed ledger that cannot be corrupted or falsified, a blockchain has Byzantine fault tolerance built into its core. A blockchain-like system allows for auditability and interoperability, and in so doing has the potential to limit the power of any one organization. Over the long term, a blockchain can financially incentivize constant cooperation, persuasion, and enfranchisement. It can create systems of cooperation that can't be beaten through a single point of failure or incompetence. Blockchain represents the real revolution the internet should have been to global power structures, because it has utterly different incentives than the existing models.

First incarnations of things are almost always laughably bad, of course, and the cryptocurrencies and blockchain solutions that dominate the conversation today are not very useful or interesting. Cryptocurrencies have become vehicles for speculation, centralization, and money laundering. Clearinghouses such as Open Sea that dominate the trade in NFTs are, at the time of writing, as centralized and impenetrable as Facebook or Google or any other tech behemoth. Cryptocurrency today is full of criminals, misfits, hackers, hustlers, and heedless triumphalists. The community will steal the NFTs right from under you just as soon as it will make you rich or sell you a rock worth thousands of dollars. But it's also full of committed idealists, and collaborative thinkers, and others who whisper persuasively of a future without a center.

Blockchain allows the construction of entities that create more value in aggregate for all participants, but do not require hyper-profitable centers that suck out all the marrow and create one-party dominance. In a world of blockchain, the big database at the heart of a tech giant becomes a service that lots of businesses can share and cooperate around. Cooperation is trivial because it's baked in, and contracts that would have been too costly to enforce are simply encoded in the public ledger. Want to share the revenue of a magical sword every time it's sold, in some complex way? No problem. Want to give rewards to other businesses that add users to a shared database? Easy. Want to build a business on another business? You can do it without discussion.

The internet is a system in which the spammers and traitorous generals almost always win. The leaders of our current data economy know this, and they have decided to pursue their own advantage no matter what doing so might mean for everyone else. But thriving in a world of blockchain requires utterly different skills in leadership than you need in order to thrive on the Web. When Ethereum or Bitcoin upgrades, those currencies must persuade the crowd to adopt change. Have you ever been polled on a change in Google's algorithm? Blockchain-like systems can incentivize transparency while disempowering corporate tyrants.

At the same time, the governance and transparency aspects of a blockchain-like system do not obviate the need for external oversight. Society must still create a framework for the conception and implementation of mechanisms that will encourage fairness, competition, and democratic governance while preventing power from consolidating in a small number of hands. These interventions must be pragmatic social ones rather than exclusively technical ones.

A FRAMEWORK FOR REGULATION WITHIN VIRTUAL WORLDS

Just as the data amassed within the internet represents a commons, the metaverse will also give rise to an enormous body of data which will power its experiences and critical services. The metaversal commons will include all of the same sort of material that accumulates on the internet, as well as the many components that will be unique to virtual worlds. On a technical level, a metaverse is a series of experiential spaces, backed up by vast amounts of data that represent content and assets of all kinds, including huge amounts of art. It is also a sequence of game-like experiences and worlds that live and run as services on infrastructure. The metaverse also consists of identities, users, transaction systems: all of the things you'd need in order to underpin a massive simulation of daily life. More than just videos, pictures, and conversations, the metaverse will be a network of vast, constantly interacting virtual worlds with trillions of interactions happening at all times. This commons will be exponentially larger than that which currently exists on the internet, which is why it is exponentially more important that it not fall into private hands.

In a growth-oriented world, regulators are often reluctant to impose strict constraints on emerging industries, lest those regulations limit the economic value of the new entities. While there is merit to this philosophy, the companies I've been talking about transitioned from scrappy new startups to world-historical powers within a few years, before many lawmakers were even able to understand how to log onto the internet, let alone why regulating it ought to fall under their purview. The data commons on which today's biggest internet companies' businesses rely achieved critical mass before governments even realized that it existed. Our existing institutions are often unable to perceive the scope of the changes that new technologies create for the world, as well as their own responsibilities to provide oversight for those changes.

Many lawmakers today are working from an analog understanding of what government's role should be in modern society. No one would seriously dispute that it's the government's job to build highways and levy taxes to fund their construction. But those same people might balk at the notion that the government should play a role in auditing Facebook or running a Facebook clone as a public utility. Government as currently constructed is not great at making rapid new decisions, or at deploying or running tech services used by millions, in large part because many lawmakers mistakenly presume that direct expertise in technology is not the domain of the state. But in a world where technology is at the heart of everything—our economy, our social order, our national defense—what use is a political class or a government incapable of dealing with these concepts or participating in running these services?

This mindset must change. In the absence of active, intelligent regulation, companies will make choices that serve their own interests at the expense of the common good. We cannot adopt the utopian mentality we had at the dawn of the internet, and just expect that enlightenment will proceed naturally from innovation. Nor can we presume that any single change, such as letting people export their personal data or having them click a button to accept tracking cookies, will magically bridge the gap between the digital dark age and some virtual paradise.

Individual empowerment within the metaverse is not as simple as packing an individual's personal data into a virtual suitcase so they can take it with them wherever they want to go. Yes, we should make it easy for people who want to take control of their own data to do so, but few users will ever bother. Similarly, it's important to be realistic about the level of complacency of the ordinary user, who is almost always seeking convenience, and who wants to see technology as an appliance. Laws that open up the technical innards of a product to support the theoretical exercise of freedoms by knowledgeable users are important; so are open-source alternatives to big commercial products. But their impact will invariably be felt most keenly in small communities of very capable users. Hobbyist hacking will never be common enough for it to be the sole answer to questions of user disempowerment, because the overall share of amateur hackers will always be just a tiny fraction of the total number of users. What we need is intelligent regulation to help keep the metaverse on course for even the most complacent users. Here are a few ideas for what that framework should entail.

Internet history teaches us that the genesis of a metaverse will probably look like this: A bunch of entrepreneurs will make a bunch of small worlds, or make attempts at creating platforms to support metaverses; these platforms might at first support only game-like experiences with relatively limited ambitions. One or two of these worlds will become, somewhat inexplicably, very, very popular. People will flock to Heist World and avoid Heist Land, even though, in many meaningful respects, they will both be broadly similar products. This is almost exactly what happened with the messaging service WhatsApp: one of many attempts at similar services that just happened to go viral due to a combination of luck and skill by its founders.

In the proto-metaverse, when one or two experience-based worlds become very popular, and their user bases grow beyond a certain point, then their developers will start to make decisions that expand a user's palette of choices and opportunities within those worlds. Then those worlds will start to become more metaverse-esque. They will offer their users opportunities to display creativity and add value, and will provide creators with opportunities to be paid for their content. This progression will not happen linearly. Complexity will emerge, as it always does, following themes that we can broadly predict, but in variations that we cannot now even begin to fathom.

I believe that excessive regulation would be counterproductive at this early stage of a metaverse's development. Instead, we should foster a safe, permissive regulatory environment in which early-stage companies and networks can operate relatively unencumbered, free to experiment and to take steps to build their communities. At a certain point, though, the most popular of these networks will start to take on massive scale. They'll reach a certain number of registered users or a certain number of in-world transactions per day, and at that point the world will no longer be just the province of hobbyists playing in a virtual sandbox. When that happens, the entities that manage the world will have assembled some large essential database used by lots of people or businesses. Regulators should in turn treat these mature worlds differently than they treat startup worlds, and we should all accept that these mature worlds have become de facto utilities.

A platform of a certain scale likely will contain within its commons an enormous amount of sensitive information. It'll store and safeguard user identities. It'll manage transactions and financial information. It'll hold custody of money and valuable assets. At that scale of network, government must make sure that these specific activities are audited and regulated. When it comes to the information, such as identity, that is absolutely essential to the ongoing safe and valuable operation of the metaverse, regulators must ensure that businesses make their systems interoperable, or at least that other businesses, when they can piggyback off those systems, are not subject to excessive restrictions and cost. Interoperability creates opportunity, while decreasing the odds that absolute power will consolidate in the hands of a few tech oligarchs.

The companies that run virtual worlds will inevitably make choices that influence their users' actions and decisions in ways that will benefit those companies. The quest to iterate an ever-more engaging virtual world leads to choices that will foster long-term engagement and encourage user retention: This trajectory is absolutely at the heart of what a metaverse business is. Companies will find ways to manipulate the behavior of their users. Government regulators must make sure that these methods are not psychologically exploitative. After a platform reaches a certain scale, its approach to behavioral modification must be subject to formal scrutiny, and audited against best practices in psychology, fulfillment, and other relevant fields.

Regulators must also work to ensure that platform providers aren't free to decide how to influence the behavior of a billion people on a whim. The implementation of algorithmic modes of behavioral modification—when those methods impact huge numbers of people—should never be exclusively left up to a platform provider, just as it's never solely a pharmaceutical company's decision to release a new drug. Regulators must weigh in first. There must be a similar regulatory body ensuring that reckless operators cannot unilaterally decide to tweak an algorithm in a way that makes a million people become unhinged. If virtual worlds will indeed be a form of health care, then governments should ensure that they do not make people sicker.

It's easier to see how to enforce real-world regulation on platform providers than it is to see how to do it for companies and workers that are wholly products of these virtual worlds. Regulating the economies that exist within virtual worlds will be a trickier proposition—but one that is just as critical. At first, employer/employee relationships within virtual worlds will be informal ones. In the early stage of a metaverse, a strict employment framework might not make sense, just as it doesn't make sense to insist on triplicate contracts and escrow accounts when you agree to buy a chair on Craigslist. The informality is what makes these transactions possible.

As worlds evolve and people start to build businesses and careers within them, though, the regulatory challenges will compound. If you're an elite cat burglar in Heist World, earning $10,000 per month, then who do you work for? Do you work for a company? Is a virtual company in the game also a real company in the real world? Who pays for health insurance? How is virtual income taxed, and who has jurisdiction to impose taxation? Which country could credibly claim oversight of a job that is performed exclusively within a virtual world incorporated in Great Britain, with servers located in Estonia, by a worker who is physically located in the United States, working for a company that isn't actually incorporated but whose founders are based in South Korea, where payment is remitted in a cryptocurrency that is based nowhere and everywhere all at once?

Existing real-world legal frameworks do not neatly map onto these new economic paradigms. At best, deploying them within the metaverse would be a sloppy kludge in the short term and an unworkable solution over the long term. We must design a new framework that allows us to intelligently differentiate between types of labor and types of obligations, and we're going to have to develop this framework well in advance of when we think we will need it.

Any plausible solution likely begins with the establishment of an empowered working group, in the spirit of the Exchange, that could devise a model of the rights and responsibilities of employing people and being employed in the metaverse. These efforts might happen within some international body or NGO, or begin with industry groups before gaining more acceptance with governments. This group must develop a framework for employment in the metaverse that learns from the tragedies of the gig economy. Virtual workers must have rights, even if those rights are not precisely identical to the ones held by those who work regular jobs in the real world; virtual employers must have some obligation to their employees that extends beyond just deeming them contractors and calling it a day. Similarly, because any metaverse is going to involve a vast number of financial transactions, basically incorporating its own banking system, we'll need to create regulatory frameworks for virtual economies. These frameworks must be put in place so that jobs and economies within this space can coexist with the real world, and that things such as taxes and accounting and record-keeping are understood and incorporated.

If all of this sounds like a whole lot of regulation and complexity, it is. And it raises another, perhaps provocative, question: Who, in the end, should hold ultimate responsibility for oversight? Should the entities that govern real-world society decide how a group of people who are really part of another society operate, think, and behave? Or does it make more sense for these virtual societies themselves to start becoming representative democracies?

The societies that will be created in the metaverse won't be common to any one nation, and because of that there will really be only two models of enfranchisement. You can build big international structures, such as the Exchange, to externally oversee and implement governance. These sorts of structures will likely be necessary in the beginning stages of virtual society—but as virtual worlds grow in both size and complexity, this solution may well prove inadequate. In time, I believe the metaverse will come to contain its own nations, and that these virtual worlds eventually should start to govern themselves.

There is nothing inherently wrong with the notion of separating, to some extent, the governance of real-world and virtual societies, and there is nothing inherently wrong with businesses organizing more like states, around communities with voting behavior, with the business aspects of these entities becoming distinct from the governance aspects. There are already legal models for companies that wish to separate their purpose from their bottom line: businesses owned by trusts, for example, or businesses that function for benefits beyond pure profit. In the United States, for instance, a business can choose to incorporate as a low-profit limited liability company, signifying that extracting every drop of financial advantage is not its sole reason for existing. My point is that we've already created structures to govern these sorts of hybrid entities that have both business and social purposes. With virtual worlds, as the process and task of administering a metaverse takes on the complexity of running a government, I believe the structures we'll end up with will come to resemble full-fledged nation-states.

On the internet, we've all just sort of accepted user disenfranchisement as a fact of life. We cannot do the same with the metaverse. We must end up in a position where people can make meaningful choices about the lives they live online. To get to this point, we'll need transparent and ethical governance guaranteed by democratic principles: elections, voting, accountability. When this transition happens, and it will, then these worlds will have no reason *not* to become their own countries. The context of these worlds will be meaningfully removed from the context of the real world, to a point where it wouldn't make sense—and wouldn't be effective—for them to exist exclusively under real-world control. These other worlds will become so meaningful that they will come to be as real as the real world for many people. This will be the first step toward metaversal speciation.

Chapter 9

ON SPECIATION

One of the first stories my parents ever taught me was Plato's allegory of the cave. In it, Plato posited that a group of people had been chained inside a cave for their whole lives, forced to stare at a blank wall on which shadows were constantly projected. To these unfortunate souls, the shadows and the cave were the entirety of their reality. Unaware of the outside world, they did not know that the shadows they saw were only projections. Their circumscribed realm was all they knew—and, within its limits, they weren't unhappy.

At a certain point, though, one of the cave-dwellers was brought to the surface, and he was overwhelmed by the intense *moreness* of the wider world. At first, he interpreted this moreness as danger, as a threat to his health and well-being. He sought to return to the safety of his world of shadows. Soon, though, his eyes acclimated to the light—and only at that point was he able to realize the limitations of his old life. While chained inside the cave, it would have been reasonable for him to think that the outside world would be just more of the same: more caves, more walls, more shadows. But the outside world wasn't just an extension of the experiences he'd previously known. It was an expansion of the entire concept of what an experience could be.

In a way, the allegory of the cave is an apt summation of the promise of the metaverse. There will come a day, not far from now, when we look back on our lives in 2022 as if we, too, had been chained to the wall of a cave, staring at flickering shadows, unable to imagine that our current diet of experiences is an artificially limited one. As I've written many times throughout this book, advanced virtual worlds linked within a metaverse won't just improve upon the sorts of experiences that we already know. These worlds will offer an entirely new set of experiences, ones that we cannot now access or really even imagine.

We are all the product of our experiences, and as the spectrum of possible experiences expands, so too will the very definition of what it means to be human. Think back to your early development, before you knew anything about math or language or logic, before you had any real lived experience. When you were small, time seemed to stretch into forever, and objectively minor stimuli might prompt extreme and extended reactions. Remember how upset you might get upon being denied a chocolate bar at a supermarket? When you were a child, the limits of your cognition and perception served to limit your world, while magnifying the import of the experiences you had therein. That chocolate bar mattered so much to you in part because, to a child, candy represented the outer bounds of joy and pleasure. You didn't just lack knowledge of the world at that stage of your life: You lacked a certain functionality that now feels like second nature. In a very real way, there was less of you.

Human development is the process of growing into a fuller way of being. Infants are not simply small adults: They literally lack such core concepts as object permanence and theory of mind. Peek-a-boo is fun because a baby does not know where you went when it can no longer see your face. The famous "rouge test" showed that it takes almost two years of life for most infants to consistently recognize themselves in a mirror. As we grow up, our worlds grow, too. Our bodies and our minds develop. Our notions of fulfillment expand. Growing into adulthood brings about an intrinsic and extrinsic moreness. Embracing this moreness is what the metaverse is really all about.

How many human lives have ever contained within them the experience of being a true hero; of sailing into uncharted waters in pursuit of life-changing adventures; of toggling between different identities with the ease of changing your shoes; of living in multiple eras or time periods? One day all of these experiences will be commonplace. In the era of virtual society, our lives will comprise an endless moreness: a dense diet of fulfilling experiences and transformative insights against which the lives we live now will seem sparse and limited. We will soon be able to stop living in the cave—the basement of the house of existence—and begin expanding outward.

The trouble, of course, is that once you exit the cave, you can never return. Though those who never leave the cave may find its confines rich and comforting, someone who has seen the rest of the world would find the cave torturous. For Plato's cave-dweller, leaving the cave didn't expand just the parameters of the outside world—it also expanded his inner world. It transformed his ability to understand, process, and perceive things.

The very act of venturing out into the world changes you forever, rendering you unable to be fulfilled by a life of artificial limits.

But what happens when the biological structures of our bodies and minds become the very limits that we yearn to transcend? At this juncture, I'd ask you to leave the safe confines of the metaverse we've been considering throughout this book. Set aside the idea of a virtual world you might access through a screen or device or even a VR headset, and instead imagine a world to which your mind can connect more directly. Imagine a world, for example, that bypasses the limitations of the human eye and injects directly into the visual cortex a vivid set of experiences that no human has yet seen or imagined. In this world, you will literally be able to see things that are now impossible.

As you'll remember from the introduction, this notion of a future in which our minds are untethered from the limits of our bodies is in fact wholly plausible. We know that the brain processes information, and we know that, in principle, this process can be wired to a machine. Crude versions of a brain-computer interface already exist. As scientists and engineers refine this technology, we will enter an era that not only offers entirely new dimensions of experience, but also offers entirely new dimensions of *you*. The future of our species is one in which the pursuit of peak fulfillment, aided by human-machine symbiosis, allows us to evolve into *many* species. Unchained to the physical bodies that have limited us for all of human history, we will finally be able to step out of the cave and embrace a fuller way of being.

It is natural to resist these changes. In the allegory of the cave, Plato suggested that the other cave-dwellers might kill anyone who tried to make them leave the cave. The cave-dwellers hated the world of "more," considering it to be *less* than their world of chains and shadows. In truth, though, they were really just afraid of what the outside world might mean for their established way of life. They worried that whatever lay outside the cave might turn out to be worse than literally being chained to a wall, forced to stare at shadow puppets for their entire lives. We cling to what we know, and sometimes have trouble believing that the unknown might be better, not worse, than our present circumstances.

As I've noted throughout this book, many of our culture's fictional stories about the digital future are morality tales that warn of the hazards of disconnecting our minds from our bodies. These stories often present virtual worlds as destructive and dangerous. Why are there so few positive stories about this future? What is the fear really about?

I think that we fear a virtual future because we worry that the inherent transmutation might somehow make us less human—which strikes me as a very arbitrary worry when juxtaposed with all the changing and becoming you already did when you grew up. Though we might sometimes look back fondly on the simplicity of childhood, most of us wouldn't choose to trade places with our infant selves. No one thinks of themselves as less human now than they were when they were an infant just because they no longer dissolve into laughter over a game of peek-a-boo. Developing new capacities isn't just a good thing: It's an evolutionary imperative on both an individual and a social level. If humanity is to continue to grow and thrive, we must overcome our fear of change and press on toward new frontiers.

OUTER SPACE, INNER EXPERIENCE

In our ongoing search for new experiential frontiers, we have long pinned our hopes on the eventual exploration and colonization of outer space. Perhaps we can attribute this vision for the future to the natural human tendency to look upward while daydreaming; perhaps it has something to do with the massive ongoing popularity of *Star Wars*, *Star Trek*, and countless other works of fiction that take as their subject the exploration of other planets and galaxies. For centuries, creatives and futurists have predicted that human curiosity and ingenuity will inevitably lead us to actual, physical other worlds somewhere beyond the stars.

While blasting off into space is a worthwhile goal, and perhaps an essential one for the long-term survival of our species, it is not the most important kind of expansion when it comes to considering the pursuit of better lives, greater happiness, more fulfillment, and a richer society. Indeed, space travel will initially lead to much more limited experiences for the first few rounds of outer-space pioneers. Though you might not realize it while casually stargazing on a dark, clear night, the universe is very sparse. It takes a long time to get anywhere; everything is spread out. Once humans do start settling other planets, their lives there will initially be pretty bleak—much like the lives of frontier settlers on Earth, many of whom sought the consolation of religion and its promises of eternal paradise to offset the miseries of a life in which their neighbors were routinely eaten by wolves. Indeed, space travelers will probably need to connect to the metaverse in order to find fulfillment!

The practical goal of space travel will likely be to find and use more resources to fuel our species' existence. The futurist Robert J. Bradbury once suggested that the best reason to travel into space would be to harness the energy of an entire star to power a massive megacomputer that could, perhaps, run infinitely complex simulations: a structure known as a Matrioshka brain. Though it might seem absurd to travel to deep outer space just to facilitate the exploration of deep inner space, there would be a certain symbiotic poetry to the process: the final frontiers of the known universe unlocking our journey into the infinite frontiers of the mind.

I understand that the vision of cruising through advanced computer simulations may not seem as adventurous as blasting off to establish colonies somewhere beyond Mars. But I would contend that there is no qualitative difference between exploring a universe bounded by natural, physical rules and exploring a simulated universe bounded by algorithmic rules devised by humans. We often depict the exploration of a Matrix-type virtual universe as a removal from the real world, but actually we can explore the real world via simulations, too. Almost anything that can happen in a physical universe could also happen within a simulated universe, as long as we know the rules of that universe. Because we can write their rules to transgress the laws of physics that constrain us here on Earth, simulated universes will also be able to offer the sorts of experiences that you *couldn't* find in a physical universe.

We already know that advanced computer programs can convincingly model aspects of reality; for those programs to convincingly model the *entirety* of reality would require no new physics, only eventual, plausible advances in computing capacity. We also know that people can derive as much or more fulfillment from virtual experiences as they take from similar real-world experiences: Remember the example from Chapter 2 of the truck drivers who like to unwind after work by playing a trucking simulator video game? If a universe we create out of computer code can be as dense with detail and meaning as the world into which we were born, and if the simulated universe can meet our needs as well as the real one can, then how and why would we observe a difference between the two?

Some people might have broad philosophical objections to existing primarily inside a simulated universe rather than a physical universe. But in order to raise those objections, they would first have to know that they're inside a simulated universe, and I'd wager that soon—sooner than you'd think—it will be hard to tell the difference. Assuming the same plausible technological advances I mentioned earlier, is it so hard to foresee a time when digital graphics can render a virtual world with the same fidelity you'd find in the real one? Isn't it possible that, at the point when you can connect your brain directly to a computer, stubbing your toe in a virtual world might activate the same pain receptors that are activated when you do so inside your own house? When we reach that point of experiential parity, then why would it even matter to you if you knew that you were in a simulation? The real/virtual binary will be a distinction without a difference.

Our fictional models for the future have occasionally acknowledged that computerized models of reality can be as or more impactful and fulfilling than the real world. The television series *Star Trek: The Next Generation* depicted a future in which the physical universe was teeming with intelligent life forms across countless inhabited planets. And yet Captain Picard and his crew still sought out the Holodeck for the sorts of adventures that the physical galaxy could not provide. Even in *Star Trek,* virtual worlds were thought to be as or more powerful and capacious than the physical worlds explored while trekking among the stars. And yet, within the show, the Holodeck was still sometimes framed as a version of Plato's cave. The ship's crew were dissuaded from spending too long in the Holodeck, from getting too obsessed with it. This miraculous space that could convincingly simulate any terrain, any experience, and any era was thought to be a potentially dangerous distraction from reality. To me, though, it's clear that the Holodeck was actually a new frontier of reality—one that was as legitimate as any planet the crew of the *Enterprise* might choose to explore.

Why does the dream of colonizing physical terrain in the known universe still feel inherently more "legitimate" than actualizing the rich, nuanced inner spaces that humans have cultivated in their minds for millennia? Our visions for the future are rooted in our narratives of the future, and those narratives fail us when they pretend that the future resembles a linear progression across a chessboard, a process in which we can understand the parameters of possibility even as we cannot predict the specific moves. In truth, the present only ever becomes the future at the point where the board morphs into an entirely different game. Space travel feels like the clear future of our species in large part because it's an extrapolation of our history thus far: What, after all, is the Space Age but the natural evolution of the Age of Sail? But you can't adequately understand the future simply by extrapolating from the possibilities available to you in the present.

Scientists and futurists aren't wrong to think that we should seek to explore the known universe. We should absolutely go to space, and we should absolutely try to visit other planets. But we should also accept the possibility that the most interesting planets we'll explore will be ones that exist only in our minds. I began this book with a prediction, and now I'm going to end it with one, too: The true future of humanity lies not just in our species bidding farewell to Earth and expanding outward into space, but also in slipping the bonds of the "real world" and expanding inward into countless strange and strangely reward-ing virtual worlds of our own design. If all of reality can plausi-bly be designed by computers and artificial intelligence, and if a simulated mode of reality can theoretically become more real than natural reality, then we owe it to ourselves and our society to delve as deeply as possible into those digital frontiers.

In the early 1920s, British explorer George Mallory was asked why he wanted to climb Mount Everest. Mallory offered a simple reply: "Because it's there." The same bravura logic will apply when it comes time to immerse ourselves into simulated realities and connect our minds to computers. Doing so won't even be a choice as much as it'll be the whole point of things: the locus of a grand, unifying human project to optimally tune the experience of being you.

INTO THE POST-HUMAN FUTURE

Plenty of people throughout history have imagined a computer-aided future for humanity. But even these great thinkers can't have understood the wild things that digital computing has now made possible. When Charles Babbage envisioned his An-alytical Engine—the first programmable mechanical com-puter—in 1837, video games and the internet weren't on his mind, let alone the prospect of connecting one's brain directly to a digital machine. We build our tools for specific reasons, and then emergent complexity inevitably takes these tools into new and fantastical directions.

The prospect of a post-human future—of connecting your-self to a computer and having your brain fully enter a virtual context—might seem alienating at first, but I would submit that

that sense of alienation is perhaps a function of the many fictional warnings against it. Part of the reason I wrote this book is because I believe our leading models of the future are too limited. They fail to grasp where our species is actually going and fail to imagine what we might become when we get there. Societal unfamiliarity with the potential power of advanced computer simulations means that our dreamers and visionaries often don't even know to aspire to the possibilities that an inner universe might present.

The prospect of human-computer symbiosis is nothing to fear. Indeed, it may well be the most desirable thing that could happen to our species. Life on a post-human plane will bring with it infinite new opportunities for happiness, expanded intelligence, and intrinsic growth on a scale that we can't even imagine today. A post-human future will create ecological value, too. Digital society will be unbelievably compact and sustainable. The energy required to simulate the human mind, at a theoretical level, is dramatically lower than the energy required to keep a human being alive in the real world. Were we to fully harness the power of a Matrioshka brain and actually build massive computers, we would literally be able to have trillions upon trillions of people living in the equivalent of luxury inside a simulation for a fraction of the energy it takes to keep them alive now.

Perhaps this scenario reminds you of *The Matrix,* and those films' visions of endless fields of humans kept alive in pods, unaware that their brains are connected to a computer simulation. But I would submit that the dystopian thing about this premise isn't that trillions of humans might live their lives wholly inside cyberspace, but that they were forced there by hostile robots that had designed the simulation as a control mechanism. I believe that, in the future, plenty of people will choose to connect directly to a simulation as a means of pursuing peak fulfillment. Why does that seem like a bad thing?

Modern-day society is not set up to encourage the pursuit of peak fulfillment. While we've largely moved on from the days of human life being characterized as "nasty, brutish, and short," we tend to perceive those who center their days around fulfillment as indulgent, weak, and innately misguided on some level. This resistance is partially a product of social conditioning, which has trained us to associate purpose with toil and character with deprivation. This narrative made sense in the Industrial Age, when economic imperatives demanded an endless supply of laborers to run the factories that produced the goods and wealth that, over the medium term, raised standards of living for everyone. Production was then a useful analogue for moreness, and, indeed, the production mentality has created many comforts. But we cannot exponentially consume resources or base our society on the fantasy that more stuff equals more fulfillment. Like the kid who eventually grows out of pining over a chocolate bar, we now must move on to more elevated pursuits and more nutritious food in order to grow as a society.

Please don't misunderstand this world of peak fulfillment as one of pure leisure—as a world devoid of drama, substance, gain, or loss. A world of peak fulfillment will be a world of even *greater* drama than the world we know now. Joy, sorrow, terror, exhilaration: All of these feelings will be magnified in the metaverse. The metaverse will be *consequential,* and within it will be mirrored every known consequential human activity—sport, culture, love, loss, war, protest, and ritual ceremony, for instance—as well as brand-new ones. Fulfillment isn't primarily about pleasure, it's about meaning, and the metaverse will create brand-new frontiers for meaning.

But the era of virtual society won't just reveal new frontiers in human fulfillment: It will reveal entirely new dimensions of existence, ones that we as a species would never otherwise be able to experience. These dimensions of experience will represent an increase of the intellectual richness of human life.

When we have worlds in which time moves differently, ones that can condense a hundred years' worth of life and experience into the span of a real-world hour; when we have worlds where people can live out their lives as geese, or as gargoyles perched atop a Gothic cathedral; when we expand the spectrum of possible life experiences out to new and unfathomable bounds, then people will live and evolve differently. Imagine aging in reverse in a virtual world; imagine a simulated universe in which humans can fly. Imagine a world where you can inhabit the body of a Galápagos tortoise and speed up time so that you can pack its entire lifespan into the hours between breakfast and lunch. With sufficiently advanced technology, we could simulate these sorts of fantastical experiences and use them not just as vectors for human fulfillment, but as the bases for entirely new ways of being.

By simulating things that could never actually be, these worlds will give us landscapes of new ideas, and opportunities to develop societies that could never otherwise exist. We'll be able to live in realities that differ from our own in every imaginary way: geography, physics, temporality. Imagine, for example, a virtual world that simulates life as it might be inside a Dalí painting: an abstract plane of distended figures in which life and time proceed according to surrealist logic. Imagine that, eventually, the technology is created that allows you to fully immerse yourself within Dalí World to live as Galatea of the Spheres. Imagine that the fulfillment you feel as a member of Dalí World's hallucinogenic society—the sensory moreness of the entire experience—exceeds the fulfillment you could ever find elsewhere. Imagine that you take this new understanding of color, shape, and dimension and use it to create great, transformative art of your own.

The social project of the metaversal era will involve a mass of people providing fulfillment and new ideas to one another, enhancing one another's lives by means of their ability to live out their differences. A society that generates more ideas and more types of people than ever before will be a good one, a just one, a strong one. Societies that encourage diversities of thought, background, intention, and identity give their members a new kind of strength. A world in which everyone is free to self-determine is one in which people are more inventive, perceptive, and productive. When we create simulated spaces that drastically expand the spectrum of possible truths to live out, then the spectrum of potential benefits for society grows as well.

And yet, as we go farther and farther down this path, the entire premise of society will change. As you come to spend more and more time within your chosen world, you will lose a certain shared context with people who have opted to exist in other worlds. Imagine that you decide to unplug, return to the "real world," and go hang out with a friend who, perhaps, spends most of his day living as an ancient Athenian inside Classics World. What will the two of you have in common? What context would you both share? If you've just spent the last hour living for a decade as a disembodied head in Dalí World, what would you even have to say to your university buddy who now spends his days in a toga at the Acropolis? True, you'd probably both have some pretty good stories to share over drinks. But the differences in your respective experiences would be so great that it might start to feel like you're speaking completely different languages.

When technology evolves to the point where virtual worlds are as or more immersive and fulfilling than the real world, then our societies will start to fragment: in language, in context, in time, in reality. This fragmentation is neither a bad thing nor a good thing so much as it's just a *thing* that will almost certainly happen regardless of how we feel about it. The era of virtual society will actually be the era of *many* virtual societies, each with its own rules, rewards, priorities, and consequences. We will splinter into hundreds of new realities, each of which will run according to its own unique logic. Eventually we will all exit the cave of the real world and step into the light of the metaverse. What happens then?

THE FRAGMENTATION OF HUMAN EXPERIENCE

I've spent a lot of time talking about the real world in this book, often in comparison to the various virtual worlds of the past, present, and future. A key component of my argument involves the ways in which virtual worlds will improve the real world—as opposed to replacing it—by creating new forms of value and meaning that we can then transfer back to Earth via the metaverse. This value will be psychological, social, and economic in nature; it will cohere, enrich, and fulfill societies and their members. I've underscored this point in part to counteract the common presumption that the rise of virtual worlds will somehow harm or destroy the real world. Let me now complicate this picture a bit.

In some ways, I've presented a false distinction between *virtual* and *real*, and this distinction will inevitably grow more meaningless as time goes by and the metaverse matures. If we accept that a simulated universe can, in theory, be indistinguishable from reality in every way that makes reality real, then we also must accept that, at that point, it will no longer make sense to observe a strict difference between "reality" and "simulation." Simulated universes will be configurable world states that actually exist, and we'll be exploring reality while we're inside them. If we can imagine the simultaneous existence of multiple simulated realities, then we must also expect that the unified context that binds us as humans will start to disintegrate, thus sending us into a fragmented future.

What makes the real world real? Our minds create our realities, and then those realities are reinforced by our social contexts. The reason why the real world seems uniquely important, at least in comparison to virtual worlds, is that we can all broadly agree on what makes the real world real. There are certain inputs of "reality" that all of our minds process and experience in roughly the same way.

Human beings from every culture and background on Earth will understand common frames of reference: the changes of the seasons, the cycles of the moon, the linear progressions of age and time, the force of gravity. The relationships between people on Earth—and between people *and* the Earth—all presume this shared context. When you travel out of town, gravity and time work the same way they did back home. (Even if you were to travel into space, gravity would work differently for you in the exact same way that it has always worked differently for everyone who has ever traveled into space.) The real world is real because it is real in the same ways for everybody. We speak fundamentally the same kind of language as the ancients, because we fundamentally have the same kind of brains.

Just as with the physical properties of reality, there are certain social contexts that most everyone on Earth understands:
land, wealth, family, health, the process of ascending within hierarchies, the ways in which humans make cultural artifacts.
Some people value some of these things more than others, but
all humans generally consider them to be meaningful. No matter who we are or where we're from, we are all bound by the
same physical rules, and we all broadly value the same sorts of
things. This shared context will begin to collapse once the metaverse starts to mature.

By now, I hope you understand my basic definition of the
metaverse: a network of meaning that links various worlds
within a set and facilitates the transfer of value between them.
In this book, I've tried to give you some intellectual tools and
points of reference with which to understand the metaverse:
what it means, why it's important, and the steps we might take
if we want to build a good one. But I want to push your thinking
about the metaverse one level further, to a place that might feel
relatively alien within the context of the "real world," and argue
that the metaverse is the first step toward the post-human, experientially capacious societies that I predicted in this book's
introduction—toward infinite societies that are infinitely different. The metaverse is the first step toward *speciation*.

Though the metaverse is, in a sense, an old idea made new by
technological opportunity, it is also the basis for an entirely new
wave of changes that our ancestors could never have fathomed.
Those changes are all roughly grouped under the notions of *speciation* and *transhumanism*. Speciation is the phenomenon of
new species emerging from evolutionary processes, while trans-

humanism, in my definition, asserts that the future cannot be represented by a one-size-fits-all vision; that it will necessarily be an expansion of possible life outcomes that will in turn bring about a parting of ways.

The metaverse is a prism, and when our shared context hits the prism it's going to refract into infinite beams going in infinite directions. If you extend the premise of an optimally valuable metaverse out to its logical conclusion, then humanity's presumed baseline of one shared reality encompassing common concerns will one day cease to exist. "Fragmentation" is a word often used in a negative context. When people think about the fragmentation of experience, they generally think about economic and social disparities: the rich and the poor, the haves and the have-nots, those who will benefit from the future and those who won't. A classic example from fiction is found in H. G. Wells's *The Time Machine*, where his Victorian protagonist travels to the year 802,701 to find society divided between two castes: the Morlocks, who live in darkness underground, and the Eloi, who live in leisure on the surface. The Morlocks, stand-ins for the working classes of Wells's time, were portrayed as vicious, ugly creatures whose labors sustained the lives of the idle, undisciplined Eloi. (The Morlocks also ate the Eloi for nutrients. There were really no winners in Wells's version of the future.)

It's important to consider the ways in which the digital divide and the fragmentation of future experience may reinforce existing social disparities or create new ones. It would be a bad thing if virtual worlds and the metaverse extended or exacerbated inequality, and in this book I think I've provided some steps we can take to avoid those negative social outcomes. But there are limits to what these sorts of contemporary mental models can tell us about the future. Wells's story was an allegorical depiction of the class divisions of his own era, not a prediction about what life would actually be like 800,000 years in the future. Even so, when we speak of the far future we often project the priorities of the present onto a society that won't even recognize them. The entire concept of *haves* and *have-nots* is rooted in the sorts of shared scarcities and contexts that will become increasingly unimportant once the metaverse expands.

The value of a thing in a virtual world depends largely on the shared context in which that thing can be said to have value. The metaverse will sort of be like that: Its participants will all value different things, and so it won't be quite so easy to make linear comparisons of value. When there are a thousand different worlds in which you can choose to live, all of which have their own notions of value—when we can actually live in these worlds and have meaningful experiences and jobs and relationships within them—then the comparative value analysis implied by the terms *haves* and *have-nots* starts to fall apart. As value and values fragment, so too does the shared context born out of common lived experience. The splintering of this context is itself a form of speciation.

Physiologically, modern humans are not substantially different from humans who lived in the medieval era. While the average person today is likely taller, healthier, and longer-lived than their medieval counterpart, both are very clearly human beings, and both would recognize the other as such if, through some time-warp accident, they were to one day meet. But that's just about all they'd recognize about each other.

Imagine if a medieval farmer were to be somehow dropped into the middle of a modern Whole Foods in Manhattan. Imagine his reaction to the sights and the smells and the general abundance; imagine his reaction to the diversity of shoppers and employees there, or the different languages he might be hearing. This "fish out of water" premise is a staple of modern entertainment, of course. But if this atemporal juxtaposition were to happen in real life, the situation wouldn't resolve as it might in fiction, with the medieval guy, like a slightly less primitive Brendan Fraser from *Encino Man,* taking a bath, buying some jeans, and sparking a romance with the pretty young checkout clerk. Indeed, the cognitive dissonance would be too extreme for there to be any easy resolution at all.

The cognitive dissonance wouldn't just derive from the medieval guy being faced with more wealth and surplus in one place than he would have encountered in his entire medieval life; it would come from being faced with a certain *moreness* of living. Whole Foods to Medieval Man wouldn't just be a bounteous display of meats and fishes and cans and greens—it would present him with a wholly unfamiliar set of experiences and identities and ideas, interrelated social premises that we take for granted but for which he would have no context.

I'm not talking about happiness here. Just because Medieval Man couldn't shop at Whole Foods doesn't mean that his life was empirically less fulfilling than our own. I'm also not talking about binary outcomes; it would be pointless to analyze this premise as one where the Whole Foods shoppers are the haves and Medieval Man is the have-not. All I mean to suggest is that the context of modern life would present an entirely different paradigm of complexity. Medieval Man's biology isn't all that different from ours, but his mental world is poorer than ours, because the outer limits of human possibility in his era were substantially more constrained than are the limits of our own. The pity you now feel toward Medieval Man and his narrow world, so too will your descendants feel about you.

Once the shared context that binds us as a species begins to fragment with the arrival of the metaversal age, we'll start to evolve in different directions. We should not presume that the mutual context humans have always shared will be enough to unite us in a future where we can all choose to live in our own preferred realities with our own preferred rules. Different people and different species have always fought over their differences, and it would be naive to think that this won't happen in a future where those differences are starker than ever. But I tend to think that this process of speciation will be a huge net positive for humanity. We must remember that a future in which we're all more fulfilled than ever may also be a world in which we are more eager to build bridges with one another. In that future, might not the drivers of today's greed be ameliorated?

Though social contexts will fracture, I believe that metaversal society will be stronger than society today. The transhuman future posits that new modes of human fulfillment brought about by immersion in worlds of ideas can be as or more important than the real world; that, rather than extending the useful lifespan of our present reality by projecting our existing structures into the future, we can sidestep existing contexts entirely and create new ones. The questions we ask ourselves about the future should focus less on what's real and more on what's important. When we reorient our thinking in that manner, a lot of the apparent problems pertaining to the metaverse will recede.

Today, pundits' worries about the metaverse are premised on a shared context we see as intractable, rather than on the fragmented contexts that the metaverse will eventually create. How will the metaverse deal with inequality? Will it be good for kids? How do we ward off cybercrime? These are fine questions to ask over the short term, but they all mistakenly presume a deathless, immutable "we."

In the metaversal future, we will face an entirely different set of practical and ethical questions, and we should start asking them now. The more disparate the context between two people, the less the process of managing people has to do with top-down governance and oversight. Instead, it becomes a process of ecological management. There will eventually no longer be a *we* to unite disparate groups of people. We will not be one species, but many; not one society, but many societies. We must account for this long-term prospect as we work to implement the metaverse and prepare to manage its short- and long-term effects.

The prospect of speciation presents a challenge for investors, regulators, creators, and anyone who is hoping to tend to the development of the metaverse. In the future, the very premise of governance will fracture. The entire idea of electing a prime minister in the context of an era in which a million constituents might be inhabiting several different worlds is a risible one. Where would the mandate for governance even come from? Whom would this minister even be able to claim to represent?

We cannot approach the metaverse thinking that it's just the next phase of the internet. Instead, we must do so fully expecting that the metaverse will transform what it means to be human. The governance structures we create and implement must be forward-thinking: made for a time where there are countless "real worlds," aware that at some point in the future the notion of the one true "real world" asserting dominance over countless virtual real worlds will feel laughable and irrelevant.

I don't fear this vision of the future, and neither should you. I believe the metaverse will make the physical world a better place, and will improve our lives—primarily by freeing us to do more, know more, be more, and experience more. Throughout human history we've sought out virtual worlds in order to expand our capacities for growth, feeling, knowledge, and relatedness. The era of virtual society will be the apotheosis of this quest. We'll be more ourselves than ever before.

In popular works of space-travel science fiction, human beings commonly coexist with other species: They work together, they live together, they mate together. This vision of the future is one that the outer-space futurists actually got right—sort of. In 1950, so the story goes, the physicist Enrico Fermi and some colleagues were discussing the prospect of alien life forms and interstellar travel. At one point, Fermi exclaimed, "But where is everybody?" If advanced alien life forms existed, then where *were* they? Why hadn't we encountered them yet?

This question became known as the Fermi Paradox. I would suggest that the extraterrestrials we thought we'd find in outer space will actually be found here on Earth. They'll be our future selves, living our lives in various different realities, evolving differently as a result of it. To Enrico Fermi, I would say: We're the aliens.

Acknowledgments

This book would not have been possible without contributions from so many colleagues, friends, and others who offered helpful commentary on early versions. However, I am especially indebted to the earliest members of Improbable: Rob Whitehead, Peter Lipka, Paul Thomas, Dima Kislov, and Jim Tang, for starting the journey that led to an obsession with the topics herein. While many other people helped build this body of knowledge, at the great risk of offending some of them through omission, I will specifically mention a few more people. Callum Lawson and Sam Snyder made specific contributions to my understanding of what is possible in virtual worlds, and these contributions are reflected in the book. John Wasilczyk and Aaryn Flynn enormously reinforced my use of self-determination theory in the book through our many chats, leading to me pursuing it as the basis for many ideas. At the Crown group of Penguin Random House, my publisher, I was very lucky to work with Paul Whitlatch, whose preternatural calm and keen editorial eye informed every stage of this process; Katie Berry, whose sharp edits made the manuscript better; Lawrence Krauser, who gave the first draft a robust and necessary copy edit; and the rest of the Crown team, including Loren Noveck, Stacey Stein, Dyana Messina, Mason Eng, Julie Cepler, Annsley Rosner, Gillian Blake, and David Drake. Finally, and most of all, I must acknowledge Justin Peters, whose partnership in bringing the book together brought the ideas to life; Marina Simon, without whom it would have been impossible to write a book while running a company; and also Laurie Erlam, Mike Harvey, and Daniel Orchard, who persuaded me to write the damn thing and worked hard to make it happen.

A Note on Sources

While I've tried my best to provide in-line references to the sources I used, there are several other texts I consulted that are worth referencing here. The following sources are ones that I found valuable while researching and writing *Virtual Society*. Some provided support for the claims I've made in this book; others served to challenge my opinions, thus forcing me to make them stronger. All are worth your time.

In **Chapter 1: Ancient Metaverses,** I consulted a wide variety of sources on comparative mythology and anthropology to shape my thinking around constructed worlds of meaning and the process by which stories can become worlds. As cited in the text, the works of Émile Durkheim, Pierre Janet, Claude Lévi-Strauss, Bronisław Malinowski, and Victor Turner were very valuable, as was J. F. Bierlein's approachable study of parallel mythmaking across human history, *Parallel Myths* (New York: Ballantine Books, 1994). And, just as a general note, there's never a bad time to read Hannah Arendt. Start with the texts I referenced within the chapter, and from there, if you're interested, move on to the following works:

- Joseph Campbell, *The Hero with a Thousand Faces* (New York: Pantheon, 1949)
- Julien d'Huy, "The Evolution of Myths" (*Scientific American,* November 2016)
- Émile Durkheim, *The Elementary Forms of the Religious Life* (New York: Free Press, 1995, originally published in 1912)
- David Gelernter, *Mirror Worlds* (New York: Oxford University Press, 1991)
- David Graeber and David Wengrow, *The Dawn of Everything* (London: Allen Lane, 2021)
- Yuval Noah Harari, *Sapiens* (New York: Random House, 2014)
- Robert Lebling, *Legends of the Fire Spirts: Jinn and Genies from Arabia to Zanzibar* (Berkeley, CA: Counterpoint, 2010)
- Claude Lévi-Strauss, *The Raw and the Cooked* (New York, Harper Torchbooks, 1964)

Chapter 2: Work, Play, and the Purpose of Free Time was influenced by David Graeber's *Bullshit Jobs* (New York: Simon & Schuster, 2018) and Daniel Markovits's *The Meritocracy Trap* (New York: Penguin, 2019), as well as a wide variety of academic papers about the evolution of labor and leisure in the Industrial Age, such as Steven Gelber's essay "A Job You Can't Lose: Work and Hobbies in the Great Depression" (*Journal of Social History,* Summer 1991), Peter Burke's essay "The Invention of Leisure in Early Modern Europe" (*Past & Present,* February 1995), and E. A. Wrigley's paper "The Process of Modernization and the Industrial Revolution in England" (*Journal of Interdisciplinary History,* Autumn 1972). Noam Chomsky's 1959 essay "A Review of B. F. Skinner's *Verbal Behavior*" (*Language* vol. 35, no. 1, 1959) helped to put behaviorism into context both for me and for the world, while Edward Deci's *Intrinsic Motivation* (New York: Springer, 1975) was valuable both in its own right and as a look at the early version of ideas he would later refine in his work with Richard Ryan on self-determination theory. For further reading:

- David Graeber, "On the Phenomenon of Bullshit Jobs: A Work Rant" (*New Poetics of Labor,* August 2013)
- Abraham Maslow, *Motivation and Personality* (New York: Harper & Brothers, 1954)
- Domènec Melé, "Understanding Humanistic Management" (*Humanistic Management Journal* vol. 1, 2016)
- Bertrand Russell, "In Praise of Idleness" (*Harper's,* October 1932)
- James Suzman, *Work: A History of How We Spend Our Time* (London: Bloomsbury, 2020)
- Frederick Taylor, *The Principles of Scientific Management* (New York: Harper, 1911)

Chapter 3: Better Experiences for Better Living is heavily indebted to the work of Edward Deci and Richard Ryan, the academic psychologists who pioneered the study of self-determination theory; their *Intrinsic Motivation and Self-Determination in Human Behavior* (New York: Plenum, 1985) is seminal. Beverley Fehr's *Friendship Processes* (New York: Sage, 1995) offered clear insights into the mechanics of making and keeping friends. In seeking to understand how and why experiences became central to human life, I very much enjoyed reading up on the history of the Grand Tour, and found Goethe's *Italian Journey* not just useful, but itself intrinsically delightful. Further reading:

- Joseph Campbell, *The Hero with a Thousand Faces* (New York: Pantheon, 1949)
- Gavin Mueller, *Breaking Things at Work: The Luddites Are Right About Why You Hate Your Job* (New York: Verso, 2021)
- Richard Ryan and Scott Rigby, *Glued to Games: How Video Games Draw Us In and Hold Us Spellbound* (New York: ABC-CLIO, 2011)
- Richard Ryan, Scott Rigby, and Andrew Przybylski, "The Motivational Pull of Video Games: A Self-Determination Theory Approach" (*Motivation and Emotion* vol. 30, 2006)
- Ben Wilson, *Empire of the Deep: The Rise and Fall of the British Navy* (London: Weidenfeld & Nicolson, 2013)

Chapter 4: A Framework for Complexity in Virtual Worlds in part examines the role that fiction has played in influencing our conception of virtual worlds throughout history. The works of Neal Stephenson and William Gibson are seminal and highly recommended. Ernest Cline's *Ready Player One* (New York: Crown, 2011) is a good read and worth your time. Chip Morningstar and F. Randall Farmer's paper "The Lessons of Lucasfilm's Habitat" (*Virtual Worlds Research,* July 2008) is very valuable, and there is much to enjoy in Andrew Groen's works that capture the ongoing history of *Eve Online*. Further reading:

- William Gibson, "Burning Chrome" (*Omni,* July 1982)
- David Karpf, "Virtual Reality Is the Rich White Kid of Technology" (*Wired,* July 2021)
- Neal Stephenson, *Snow Crash* (New York: Bantam, 1992)
- Rob Whitehead, "Intimacy at Scale: Building an Architecture

for Density" (*Improbable Multiplayer Services,* June 1, 2021,
ims.improbable.io/insights/intimacy-at-scale-building-an
-architecture-for-density)

Chapters 5 through 9, being inherently forward-looking,
relied less on research than did the preceding chapters. Instead,
the ideas found in this section of the book are rooted in my own
experiences and insights as the co-founder of a company that
makes virtual worlds, and in my constant conversations with
colleagues, scholars, and industry leaders about the opportunities and challenges of the coming metaversal era. That said, I
also consulted many texts that helped hone and challenge my
own ideas. The works cited below all played a part in helping to
bring these chapters to life.

CHAPTER 5: A NETWORK OF MEANING

- Acceleration Studies Foundation, "The Metaverse Roadmap"
  (2007, www.metaverseroadmap.org/overview/)
- Matthew Ball, "The Metaverse Primer" (June 2021,
  www.matthewball.vc/the-metaverse-primer)
- Edward Castronova and Vili Lehdonvirta, *Virtual Economies:
  Design and Analysis* (Cambridge, MA: MIT Press, 2014)
- Nikolai Kardashev, "Transmission of Information by Extrater-
  restrial Civilizations" (*Soviet Astronomy,* September–October
  1964)
- Raph Koster, "Still Logged In: What AR and VR Can
  Learn from MMOs" (GDC talk, 2017, www.youtube.com/
  watch?v=kgw8RLHv1j4)
- Kim Nevelsteen, "A Metaverse Definition Using Grounded
  Theory" (September 2, 2021, kim.nevelsteen.com/2021/09/02/
  a-metaverse-definition-using-grounded-theory/)

CHAPTER 6: BUILDING A VALUABLE METAVERSE:
THE EXCHANGE

- Raph Koster, "Riffs by Raph: How Virtual Worlds Work" (*Playable Worlds,* September 2021, www.playableworlds.com/news/riffs-by-raph:-how-virtual-worlds-work-part-1/)

- Carl David Mildenberger, "Virtual World Order: The Economics and Organizations of Virtual Pirates" (Public Choice vol. 164, no. 3, August 2015)

- Dan Olson, *Line Goes Up—The Problem With NFTs* (video essay, January 21, 2022, www.youtube.com/watch?v=YQ_xWvX1n9g)

- Camila Russo, *The Infinite Machine: How an Army of Crypto-hackers Is Building the Next Internet with Ethereum* (New York: HarperCollins, 2020)

- Laura Shin, *The Cryptopians: Idealism, Greed, Lies, and the Making of the First Big Cryptocurrency Craze* (New York: PublicAffairs, 2022)

CHAPTER 7: VIRTUAL JOBS AND THE
FULFILLMENT ECONOMY

- Edward Castronova, "Virtual Worlds: A First-Hand Account of Market and Society on the Cyberian Frontier" (CESifo Working Paper No. 618, December 2001)

- Kei Kreutler, "A Prehistory of DAOs: Cooperatives, Gaming Guilds, and the Networks to Come" (*Gnosis Guild,* July 21, 2021, gnosisguild.mirror.xyz/t4F5rItMw4-mlpLZf5JQhElbDfQ2JRVKAzEpanyxW1Q)

- John Rawls, *A Theory of Justice* (Cambridge, MA: Belknap Press, 1971)

CHAPTER 8: THE TYRANTS AND THE COMMONS

- Finn Brunton, *Spam: A Shadow History of the Internet* (Cambridge, MA: MIT Press, 2013)
- Julian Dibbell, "A Rape in Cyberspace, or How an Evil Clown, a Haitian Trickster Spirit, Two Wizards, and a Cast of Dozens Turned a Database into a Society" (*Village Voice,* December 21, 1993)
- Chris Dixon, "Why Decentralization Matters" (*One Zero,* February 18, 2018, onezero.medium.com/why-decentralization -matters-5e3f79f7638e)
- Niall Ferguson, *The Square and the Tower* (New York: Penguin Books, 2018)
- Milton Friedman, *Capitalism and Freedom* (Chicago: University of Chicago Press, 1962)
- Scott Galloway, *The Four: The Hidden DNA of Amazon, Apple, Facebook, and Google* (New York: Portfolio/Penguin, 2017)
- Katie Hafner and Matthew Lyon, *Where Wizards Stay Up Late: The Origins of the Internet* (New York: Simon & Schuster, 1996)
- Leslie Lamport, Robert Shostak, and Marshall Pease, "The Byzantine Generals Problem" (*ACM Transactions on Programming Languages and Systems,* July 1982)
- Steven Levy, *Facebook: The Inside Story* (New York: Blue Rider Press, 2020)
- Carl D. Mildenberger, "The Constitutional Political Economy of Virtual Worlds" (*Constitutional Political Economy* vol. 24, no. 3, September 2013)
- Thomas More, *Utopia* (1516)
- Justin Peters, *The Idealist: Aaron Swartz and the Rise of Free Culture on the Internet* (New York: Scribner, 2016)
- Matt Stoller, *Goliath: The 100-Year War Between Monopoly Power and Democracy* (New York: Simon & Schuster, 2019)

CHAPTER 9: ON SPECIATION

- Aaron Bastani, *Fully Automated Luxury Communism* (London: Verso Books, 2019)
- Nick Bostrom, *Superintelligence* (Oxford, UK: Oxford University Press, 2014)
- Robert J. Bradbury, "Matrioshka Brains" (1997, www.gwern.net/docs/ai/1999-bradbury-matrioshkabrains.pdf)
- David Eagleman, *Livewired: The Inside Story of the Ever-Changing Brain* (New York: Pantheon, 2020)
- Max Tegmark, *Life 3.0* (New York: Knopf, 2017)
- H. G. Wells, *The Time Machine* (1995)

Index

PENGUIN PARTNERSHIPS

Penguin Partnerships is the Creative Sales and Promotions team at Penguin Random House. We have a long history of working with clients on a wide variety of briefs, specializing in brand promotions, bespoke publishing and retail exclusives, plus corporate, entertainment and media partnerships.

We can respond quickly to briefs and specialize in repurposing books and content for sales promotions, for use as incentives and retail exclusives as well as creating content for new books in collaboration with our partners as part of branded book relationships.

Equally if you'd simply like to buy a bulk quantity of one of our existing books at a special discount, we can help with that too. Our books can make excellent corporate or employee gifts.

Special editions, including personalized covers, excerpts of existing books or books with corporate logos can be created in large quantities for special needs.

We can work within your budget to deliver whatever you want, however you want it.

For more information, please contact
salesenquiries@penguinrandomhouse.co.uk